CHOOSING EARTH

Humanity's Journey of Initiation *Through* Breakdown
and Collapse to Mature Planetary Community

Revised Edition

Duane Elgin

Preface by Francis Weller

Published by Duane Elgin
www.DuaneElgin.com
Copyright © 2022 Duane Elgin

This book is part of the Choosing Earth Project
For more information: www.ChoosingEarth.org

All rights reserved. No part of this book may be reproduced or modified in any form, including photocopying, recording, or by any information storage and retrieval system, without written permission from the publisher, except as provided by the United States of America copyright law. For permission requests, write to Duane Elgin through www.ChoosingEarth.org.

Book design and infographics: Birgit Wick, www.WickDesignStudio.com
Karen Preuss cover photographer
Graph page 56: Emily Calvanese
Font: Georgia and Avenir Next
First Edition: March 2020
Second Edition: January 2022
ISBN 978-1-7348121-3-8

Endorsements for *Choosing Earth*

"Duane Elgin's masterpiece is the most powerful and comprehensive wake-up call on Earth ... a passionate, eloquent, and wise book."
—Alexander Schieffer, professor and co-author of *Integral Development*.

"I have never before read a book on the global climate crisis by a "white American male" that has so deeply touched and enriched me."
—Rama Mani, PhD, World Future Council, convenor and organizer.

"*Choosing Earth* provides a bold and hopeful vision of the next 'holistic' stage of human civilization."
—Bruce Lipton, PhD, biologist, speaker, author of *Biology of Belief*.

"We humans have a third choice—to respect ecological boundaries and regenerate the Earth for the well-being of all."
—Vandana Shiva, environmental activist, scholar, author of *Earth Democracy*.

"Duane Elgin has done the hardest work that none of us ever want to do. Reading Choosing Earth *will change you forever.*"
—Sandy Wiggins, green building, mindful business, ecological economics.

"Your excellent book is very much in line with our concerns and priorities. My warmest personal regards."
—Antonio Guterres, Secretary-General of the United Nations.

"*Choosing Earth* describes the only possibly viable path ahead—a tumultuous path of initiation to full maturity as members of the living world."
—Eric Utne, founder, Utne Reader, author of *Far Out Man*.

"This is one of the most important books for our time—and probably the most important document on the perils of climate change. Every politician and CEO needs to read this."
—Christian de Quincey, philosopher, author of *Radical Nature*, teacher.

"Duane Elgin's panoramic wisdom in Choosing Earth *is vital in this time when complex, interconnected crises demand coherent, interconnections solutions. A pioneering and important book.*"
—Kurt Johnson, PhD, biologist, Inter-spiritual leader, professor, author.

"All life on Earth owes a debt of gratitude to Duane for awakening us to the urgency and regenerative possibilities of Choosing Earth."
—John Fullerton, former managing director at JP Morgan, founder of Capital Institute.

Dedicated to

Roger and Brenda Gibson
Whose soulful and financial support
Launched the Choosing Earth Project

and

Coleen LeDrew Elgin
Whose love, partnership, and tireless efforts
Brought this work into the world.

Contents

Preface: At the Threshold: Grief, Initiation, and Transformation
by Francis Weller 7

PART I: OUR WORLD IN GREAT TRANSITION
Humanity's Initiation and Transformation 24
Growing Resilience in a Transforming World 30

PART II: THREE FUTURES FOR HUMANITY
Extinction, Authoritarianism, Transformation 42
Future I: Extinction 47
Future II: Authoritarianism 51
Future III: Transformation 52

PART III: STAGES OF INITIATION AND TRANSFORMATION
Scenario of Transformation 60
2020s: The Great Unraveling—Breakdown 63
2030s: The Great Collapse—Free Fall 81
2040s: The Great Initiation—Sorrow 89
2050s: The Great Transition—Early Adulthood 101
2060s: The Great Freedom—Choosing Earth 109
2070s: The Great Journey—An Open Future 115

PART IV: UPLIFTS FOR A TRANSFORMING FUTURE
Uplifts for Transformation 118
Choosing Aliveness 118
Choosing Consciousness 134
Choosing Communication 139
Choosing Maturity 148
Choosing Reconciliation 152
Choosing Community 161
Choosing Simplicity 165
Choosing Our Future 174

Acknowledgements 177
A Personal Journey 179
Endnotes 184

PREFACE

At the Threshold: Grief, Initiation, and Transformation

by Francis Weller

"In a dark time, the eye begins to see."

—Theodore Roethke

We are living in turbulent times on this beautiful planet. All pretense of immunity is collapsing as we realize how completely entangled our lives are with one another—with kelp beds and calving glaciers, with wildfires and rising sea levels, with refugees and the anxious dreams of young people everywhere. The disequilibrium shaking the world feels like a continual tremor along the fault lines of our psychic lives.

Very few things feel stable. It is like a fever dream. Maybe we have reached the initiatory threshold required to wake us up. Whatever is happening, much will be asked of us if we are to make it through the whitewater of this narrow passage. We do not know what lies ahead, but one thing is sure: *This is a time for bold gestures.* It is time to wake up and humbly take our place on this stunning planet. The future is speaking ruthlessly through us.

James Hillman, the brilliant archetypal psychologist, wrote, "The world and the gods are dead or alive according to the condition of our souls."[1] In other words, the vitality of the animate, sensuous world and our encounter with the sacred depend on our souls being fully alive! A soul that is awake is entangled with the living world—its beauty, allure, and wonder, its sorrows, rips, and tears. Given the state of the world and our soulful lives, we must pause and ask, "*What is the condition of our souls?*" From all observable accounts, the prevailing condition is desperate, empty, ravenous, impoverished, and grief-stricken. In the language of some traditional cultures, we would diagnose our times as one of *soul loss*. To lose soul is to feel emptied of wonder, joy, and passion. It is to feel cut off from the vitalizing relationships with the living world, leaving one stranded in a deadened world. The long-standing intimacy with the multiple folds of the Earth—her myriad of creatures, the stunning profusion of color and fragrance—would be forgotten. In place of this, we substitute a frenzied striving for power and material gain. This is the dominant reality for much of white, technological, late-capitalistic culture. Soul loss leaves us flattened and empty, always wanting more—more power, more things, more wealth, more control. We forget what truly satisfies the soul.

I have spent nearly four decades tracking the movements of soul, most especially through the layers of grief. In my practice as a psychotherapist and in many workshops, I have seen the wide range of sorrows that we carry in our hearts. From early traumas, deaths, divorces, suicides of beloved family or friends, addictions, illnesses, and more . . . the "size of the cloth" has become painfully apparent. More and more frequently, I hear in the laments of individuals, not so much grief for their personal losses, but for the wider, wilder world that is being diminished minute by minute. They are registering in their souls the sorrows of the world. Strangely, this gives me hope.

The sheer weight of these personal and collective sorrows is enough to crush our hearts, forcing us to turn away and find solace in anesthesia and distraction. When we come together, however, and share these stories of sorrow in grief rituals, something begins to change. When our sorrows are witnessed and held within a community of compassion, grief can surprisingly turn to joy, to a love emboldened for all that surrounds us. Love and loss have been eternally entwined. To acknowledge our grief is to free our love to fall outwards into the waiting world.

Something *is* stirring in the depths of the times. Our collective denial appears to be cracking. We can no longer deny the fact that the world is radically changing. We sense in our bones the breakdowns occurring and, along with it, our hearts feel weighted with grief. It may be our shared sorrows, stirred by our love of this singular, irreplaceable planet, that will ultimately activate our communal commitment to respond to the rampant denigration of the world. Robin Wall Kimmerer writes, "If grief can be a doorway to love, then let us all weep for the world we are breaking apart so we can love it back to wholeness again."[2]

The Long Dark

Duane Elgin's *Choosing Earth* is a demanding book, asking us to do the hard work of turning into the coming waves of breakdown, bewilderment, chaos, and loss. He invites us to participate in the most difficult transition humanity will ever have to make—an invitation we hoped never to receive. Its arrival declares that the planet has already radically and irreversibly changed and it is now up to us to respond. Yet, hidden within this ominous threshold-time are the seeds of humanity's possible maturation into a planetary community. As this book lays bare, however, the passage will be long, and we will be working these evolutionary changes for decades and, most likely, for generations to come. So, dear reader, persist, even though it is difficult. Even though your heart breaks a thousand times. As Buddhist scholar and eco-philosopher Joanna Macy said, "The heart that breaks open can contain the whole universe."

Elgin does not offer prescriptions for fixing what is happening, nor encourage some return to a better past, nor does he suggest we surrender to ruin. He soulfully recognizes that we must go *through* this time of collective initiation in order to take our place as responsible adults collaborating in the creation of a healthy and vibrant community of all beings. This is challenging reading. Much will be evoked as you take in the information, the timelines, and the grief of our evolving story. Read on. The future is not set and each of us is a makeweight in the shaping of what is to come.

This descent takes us down into a different geography. In this shadowed terrain, we encounter a landscape familiar to soul—loss, grief, death, vulnerability, fear. This is a time of decay, of shedding and endings, of falling apart, and collapse. This is not a time of rising and growth. It is not a time of confidence and ease. No. We are hunkered down. "Down" being the operative word. *From the perspective of soul, down is holy ground.* We are being escorted into the hallways of soul.

We are entering what could be called the *Long Dark*. I say this not with a note of despair, or with an attitude of hopelessness, but, instead, recognizing and valuing the necessary work that can take place only in the dark. It is the realm of soul—of whispers and dreams, mystery and imagination, death and ancestors. It is an essential territory, both inevitable and required, offering a form of soul gestation that gradually gives shape to our deeper lives. Certain things can happen only in this grotto of darkness. Think of the wild network of roots and microbes, mycelium, and minerals, making possible all that we see in the day world, or the extensive networks within our own bodies, bringing blood, nutrients, oxygen, and thought to our corporeal lives. All of it happening in the darkness.

Collectively, we are not familiar with descent as something valued and essential. Most of us live in an ascension culture. We love things rising up ... up ... up ... always up. When things begin to go down, we can feel panic, uncertainty, and even dread. How can we meet these unpredictable times with any sense of courage and faith? Courage to keep our hearts open and faith that something meaningful lurks in the descent. How can we, once again, come to see the holiness that dwells in darkness?

To remember the sacredness in the dark, we must become fluent in the manners and ways of soul. We are required to develop another way of seeing as we descend ever-further into the collective unknown. We are asked to remember the disciplines of soul that will enable us to navigate through the *Long Dark*. This is a time to practice *deep listening,* which acknowledges the wisdom in others and in the dreaming Earth. When we listen deeply, we begin to uncover what wants to be brought into being. As Alexis Pauline Gumbs, a black feminist writer and poet, asks, "How can we listen across species, across extinction, across harm?"[3]

Qualities and disciplines we need to collectively practice include the following:

- *Restraint* offers a breath, a pause, a moment of reflection, which allows things to be revealed. Restraint enables something to ripen before we move into action.
- *Humility* honors our mutuality and brings us close to the ground, a gesture that keeps us aware of our entanglement with the living world.
- *Not knowing* reminds us that we live in mystery, an ever-unfolding, unshaped moment. We do not know what is going to happen, and this truth keeps us humble and vulnerable. And finally . . .
- *Letting go* . . . rooted in the fundamental truth of impermanence. Each of us is preparing for our own disappearance as well as witnessing the constantly shifting world. We are reminded of the continual process of change.

Each of these disciplines helps us to cultivate our presence in the underworld of the *Long Dark*. Primary among the skills we need to cultivate in these uncertain times is our capacity to grieve. Even our basic trust in the future has been shaken as we awaken to the emerging climate crisis and erosion of the social fabric. As a result, we now face a vital truth: We are entering a *rough initiation*.

Rough Initiations

Uncertainty has come into our homes and found its way into each of our lives. What was once stable and predictable has been shaken and we have begun a steep descent into the unknown, surrounded by insecurity, fear, and grief. Many of my clients confess that what troubles them most is the condition of the world! The symptoms are no longer confined to our intrapsychic realities—our personal histories, wounds, and traumas. The patient is now the planet itself, manifesting symptoms of collapse, depression, anxiety, violence, and addiction—felt in the wider body of the Earth, rattling our deep psychic ground, affecting everything.

Hidden within our shared experience of suffering, are the unripened seeds of initiation.

Daily, we receive news of the latest frightening climate report, of violations to our human and more-than-human kin, of tragedies in all parts of the world. Our psyches are inundated. The scale of suffering and loss is hard for us to comprehend as individuals. We are not wired for this level of persistent, collective trauma. We are designed to metabolize the challenges and sorrows of our local community and our own encounters with suffering. Learning to digest this wider emerging reality requires the support of community, rituals that can help us stay connected to our souls, along with a compelling story that invites us to dream of what is possible. Without such deep connections, we will continue to rely on strategies of avoidance and heroic striving, hoping to bypass painful encounters.

As we slowly digest the contents of *Choosing Earth*, we come to realize that we are tumbling through a *rough initiation,* with radical alterations occurring in our inner and outer landscapes—simultaneously deeply personal and wildly collective, binding us to one another. Everyone we meet—in the grocery store, in line at the gas station, walking their dog—is tangled up in this liminal space between the familiar world and the strange, emergent one. Hang on!

The deep work of traditional initiations was meant to dislodge an old identity. The process was designed to produce enough intensity and heat to cook the soul and prepare initiates to take their place in the care and maintenance of the commons. *It was never about the individual.* It was not about self-improvement or making them into someone better. No. *Initiation was an act of sacrifice on behalf of the greater community into which the initiate was brought and to which he or she now holds allegiance.* They were being readied to step into their role of maintaining the vitality and well-being of the village, the clan, the watershed, the ancestors, and the continuum of generations to come.

We are meant to be radically changed by initiatory encounters. We do not want to come out of these turbulent times the same as we went in, personally or collectively. At this moment in history, we need to respond to *radical change.* This period of rough initiation has been brought about by multiple crises: economic instability, cultural and political upheavals, massive relocations of refugees, racial and gender injustice, food and water shortages, uncertain availability of healthcare, and others. Undergirding them all is the collapse of our ecological systems. As this reality comes closer and our imagined separation from nature is thinned, we recognize that our sense of who we are is entirely entangled with coral reefs and monarch butterflies, blue fin tuna, and old-growth forests. Their decline is our diminishment. As Elgin writes, "Eco-collapse brings ego collapse." The Earth container is breaking, and with it the fiction of separation. Our rough initiation is bringing about the death of our collective adolescent identity. It is time to ripen.

So now what? How do we navigate this tidal surge of uncertainty? How do we engage the world in the absence of the ordinary? Fear can rattle us and activate strategic patterns of survival. This is evident in the resurgence of old modes—such as scapegoating, projection, hatred, and violence. These patterns may allow some to temporarily avoid the descent, but those strategies cannot help us across this tremulous threshold into a planetary civilization. For that, we need to amplify the potency of the adult. As is true of any genuine initiation, it requires a maturation of our being and stepping more fully into our robust identity, rooted in soul. We must become immense, capable of welcoming all that arrives at the gateway to the heart.

An Apprenticeship With Sorrow

Our collective initiation will inevitably bring us face-to-face with extreme layers of loss and grief. Elgin makes this very clear. The ongoing winnowing of species will deplete the Earth's biodiversity by a staggering amount over the coming decades. Human deaths will multiply as food and water sources disappear and regional violence increases over diminished access to resources. Economic disparities will level an untold degree of suffering on billions of individuals. *Grief will be the keynote for the foreseeable future.* Our ability to stay present to this tidal wave of loss depends on our capacity to cultivate this essential skill. We must take up an *apprenticeship with sorrow.*

Our apprenticeship begins when we come to understand that grief is ever-present in our lives. This is a difficult realization, but one that has the opportunity of opening our heart to a deeper love for our singular life and for the wind-swept world of which we are a part. We begin with the simple gesture of picking up the shards of grief that lie littered on the floor of our house. We begin by building our capacity to hold sorrow in the tender hut of the heart. Through this practice, we learn to welcome the pervasive and encompassing presence of grief. And then we invite one, two . . . a few trusted others, to gather and share the ongoing waves of sorrow as they come ashore. "Our ability to love and comfort is expanded by others' grief, our own too-big-to-be-contained pain finds its freedom in others' witnessing of it."[4]

Grief is more than an emotion; it is also a core *faculty of being human.* It is a skill that must be developed, or we will find ourselves migrating to the margins of our lives in hopes of avoiding inevitable entanglements with loss. Through the rites of grief, we are ripened as human beings. Grief invites gravity and depth into the psyche. Fortunately, we possess the capacity to metabolize sorrow into something medicinal for our soul and the soul of the world.

One of the essential practices in our apprenticeship is our ability to hold one another in times of grief and trauma. This skill has, for the most part, been lost under the extreme weight of individualism and privatization, especially in Western, industrial cultures. This has had a profound impact on how we process and metabolize our personal encounters with loss and intense emotional experiences. Without the familiar and reliable container of community, these times can penetrate our psychic lives, leaving us shattered and shaken, frightened and unsure of our next step.

Trauma is any encounter, acute or prolonged, that overwhelms the capacity of the psyche to process the experience.

In these times, what confronts us is too intense to hold, integrate, or comprehend. The emotional charge saturates our capacity to make sense of the experience, and we feel overwhelmed and alone. Absence of an adequate holding environment, capable of supporting us in these times, generates traumatic experiences. In other words, pain itself is not traumatic. *Unwitnessed pain is.* This time of rapid and heartbreaking planetary change reminds us that we are in this together and we can offer one another the holding space needed to process our shared sorrows.

But what of traumas impinging on us from the wider world? Here, Elgin proposes a new way of framing the global field. He introduces *Chronic Planetary Traumatic Stress,* and writes: "The difference between PTSD (post-traumatic stress disorder) and CPTS is that, instead of a relatively brief and confined episode, the trauma is life-long and planetary in scope. There is no escape—the burden of collective trauma permeates the psyche and soul of humanity." There is no escape! Whether we acknowledge the wider traumas or not, our psyches register the disruption. How could we not? Our lives, our bodies, our souls, are entirely entwined with the beauty and sorrows of the world. As Elgin points out, without containment, the chronic traumas of the planet will leave many of us "deeply wounded, both psychologically and socially." The capacity to create spaces potent enough to hold the intense energies of our raw grief is a key element in our apprenticeship with sorrow.

Every trauma carries grief within it. Loss is woven within trauma's textures; and the scenarios laid out by Elgin for the next decades and beyond are filled with trauma and sorrow.

How are we to respond when life confronts us with overwhelming circumstances? How can we hold all that we feel when the source is far beyond our ability to control? How do we recalibrate our inner lives to heal our psyches in times of trauma? Here are a few offerings for tending our souls during traumatic times—and who isn't living in traumatic times?

1. **Practice self-compassion**. Self-compassion helps us hold our vulnerability with kindness and tenderness, allowing us to remain soft and open. Times of great uncertainty call for a level of generosity to ourselves that helps offset the effects of trauma that can often envelop our emotional body. This must be our first and primary intention: to hold all that we experience with compassion—to offer a safe place for our fears and grief to land.

2. **Turn toward the feelings**. No bypass or strategy of avoidance can help resolve the difficult emotions we will encounter. Turning toward our suffering is essential. Not only must we endure times of pain and sorrow, hoping to get beyond them, we must also actively engage them and feel them fully. This move takes great courage. However, without adequate compassion and support, it is hard to open ourselves up to the painful emotions that await us.

3. **Be Astonished by Beauty**. Trauma has a profound impact on our feelings of aliveness, often generating a state of numbness or anesthesia. This anesthetized state protects us for a time from having to encounter the raw, searing emotions that often accompany trauma, but it also dulls our sensual engagement with all that surrounds us. Beauty's allure helps to fully open the aperture of the heart. Sorrow and beauty side-by-side. The soul has a fundamental need for encounters with beauty—a central source of nourishment that continually renews our sense of vitality and awe.

4. **Patience**. Healing from trauma takes time. Patience helps heal vulnerable pieces of soul splintered by trauma. Knitting a

bone takes time. Mending the soul takes even longer. Be patient with your process. The soul's deep wisdom knows the value of going slowly. Stepping out of the manic pace of modern culture is essential for regaining our footing in the world of soul. Patience is a discipline, a practice that reassures wounded, vulnerable souls, and helps us reap the benefits of our efforts.

A Gradual Awakening, An Emerging World

Our long apprenticeship with sorrow results in a spaciousness capable of holding it all—the loss and the beauty, the despair and the longing, the fear and the love. *We become immense.* Our steady devotion to working with the heavy cargo of grief, slowly softens the heart and we feel our connection with the wider, sentient world expanding. Our time in the depths helps us to develop a felt intimacy with the Earth and the cosmos. We come home. We sense a diminishing distance between us and others. Our identities become permeable, and we feel a growing kinship with the human and more-than-human community. A new reverence for life emerges as we sense the living presence of the Earth as an organism embedded in a living cosmos.

This is our dawning experience of a possible future for the Earth. A mature humanity is emerging, but it is tender, vulnerable, and fragile. We are entering our early adulthood, not yet developed sufficiently to withstand much pressure. Thresholds are tenuous, unsteady, and unpredictable. As we enter, what Elgin calls "The Great Transition," we are required to return to humility again and again. What humanity has endured over the *Long Dark* must now be harvested with patience. Our work is to protect this emerging sensitivity and pass it along to the generations that follow. Each successive generation can fortify this evolving awareness, adding its own understandings, practices, rituals, songs, stories, and more—until it becomes a robust presence in accord with the evolving cosmos.

As we mature as a species, we enter a more reciprocal relationship with the Earth. We are called to strengthen the values and practices that help sustain the body of this exquisite world. Values such as respect, restraint, gratitude, and courage help to fortify our ability to stand and protect what we love. Reverence and humility remind us that our lives comingle with all life. What affects one strand on the web affects all. We are here to participate in the ongoing creation, to offer our imagination, affection, and devotion to sustaining the world.

Elgin makes the need clear: we must cultivate a robust collective of adults whose primary fealty is to the life-giving world on which we depend. We must be able to feel our loyalties to watersheds, migratory pathways, marginalized communities, and the soul of the world. We must feel the bedrock of our aliveness, and the reality of our wild and exuberant lives. Initiation tempers the soul, drawing out its hidden essence and calls forth the medicine we came to offer this stunning world. We are needed!

Initiation ripens us and readies us for greater participation in the care of the cosmos. This is at the heart of why we are here as a species. Our cosmological purpose is to keep the dream of the world alive. There is beauty, dignity, and grandeur in that calling. It is becoming increasingly clear that this realization must become deeply embedded in the hearts and souls of people in the coming decades. In essence, we are being asked to consecrate our lives, to practice reverence in our actions. This is the first truth that must settle into the bones of anyone who undergoes this planetary initiation. In addition, initiation implies soul medicine. We are asked to give away the particular gifts we came here to offer. Initiation also loosens the tight collar of civilization and leads us to reclaim the wildness within. The grip on our domesticated psyche relaxes and we are able to enter a multicentric world where everything possesses soul and is a form of speaking. And one last truth that comes with initiation: We are asked to build a house of belonging that can extend places of welcome to those who feel unseen and disconnected.

For those of us privileged with the gift of advanced years, it is incumbent upon us to turn and face those who follow, the generations of younger ones whose future is seriously jeopardized by our neglect of the world. I see the understandably bewildered, angered, grief-stricken faces of millions. I don't know what to say, only that I see you. I acknowledge your sorrow and your despair, your outrage, and confusion. Your trust in any possible future is being eroded day-by-day. What you inherently expected—a future brimming with possibility—is fading and evaporating even as you reach for it. I feel the immense sorrow in your hearts. I see it whenever we share a moment. It is etched on your face, in your words. I am sorry. Please know that many of us are doing everything we can to find a way through this narrow passage to offer a world worthy of your lives.

I also see your passion and your commitment to fight for a life that has meaning and beauty, belonging, and joy. I see your longing to fashion a living culture congruent with the ways and rhythms of the Earth. I see your creativity and wild imaginings, seeing things in ways my generation never dreamed of. You are powerful in the midst of your grief. You have been asked to carry so much, so soon, and the initiatory impulse may have been activated before you were ready. And maybe not. You may be the ones capable of finding a pathway through this collective dark night of the soul.

A New Human, A New Earth

It is a privilege to be alive at this moment in our collective story. We are the ones straddling this threshold-time. We are the ones who can choose to participate in the repair of the Earth and in the creation of a living planetary culture. We are the ones alive at a moment of immense possibility when we can restore a sacred covenant with the animate world. We are the ones who can respond to these circumstances and participate in imagining the shape of a new Earth. The Earth, however, is profoundly wounded and will require patient restoration. Attending to the sacred duty of repair is a deep imprint of our initiation.

Every human being alive will experience the rough initiation of these times. No one will be exempt from the effects of the deteriorating climate or the stresses and strains that will befall our economic, political, and social lives.

Initiation is not optional. The lingering question is will we choose to participate in the process of initiation? Will we be able to see beyond self-interest and be capable of *thinking like a planetary community?* We will be reshaped in profound ways, one way or another. If we choose to accept the challenges of this threshold-time, we may emerge ripened and ready to participate in, what geologian Thomas Berry called, *the dream of the Earth*. The hallmarks of this new self will reveal someone *more attuned to responsibilities than rights, more aware of multiple entanglements than entitlements*. We will be initiated into a vast sea of intimacies, with the village, star clusters, and gnarled old oaks, wide-eyed children, the pool of ancestors, and the scented Earth.

The importance of this choice cannot be overstated. By participating in the difficult work of radical change, we are quickened, in some deep way, to carry essential medicines for our beleaguered world. This implies that we learn how to live within the means of the Earth to support us.

"Choosing Earth" means choosing simplicity, community, reconciliation, and participation. These are gestures we can all make, now. We can remember our *primary satisfactions*—the core constituents of a healthy soul life. These elements evolved over several hundred-thousand years and shaped our psychic lives in ways that led to a sense of contentment and satisfaction. When these requirements are met, we do not crave the newest device, or the latest model car, or the next form of anesthesia. Essentially, we are freed from toxic consumerism and materialism. We live simply and we simply live. In order to feel satisfied, we need touch that affirms and soothes, to be held in times of grief and pain; we also need abundant play, and sharing food with others, eaten slowly over heartfelt conversations; we need dark, starlit nights when words are not required; and, of course, we need the pleasures of friendship and unselfconscious laughter.

We require a vital ritual life that connects us with the unseen world in crucial times—such as crossing the threshold of initiation, tending to the vulnerabilities of illness, or celebrating our communal gratitude for the blessings of this life. We need an ongoing, intimate, and sensual connection to the wild pulse of nature; our hearts and ears need to delight in storytelling, dancing, and music. We crave the attention of engaged elders, and we thrive in a community rooted in a system of inclusion based on equality. These are what we truly desire.

Let us be willing to descend, together, into the vast darkness of this time and see what resides there, in the mystery, waiting for our devoted attention. So much is in bud, the poet says. So much yearning for expression. A larger journey lies ahead, one where we may find ourselves growing into something unimagined, birthing a new being, a bio-cosmic presence.

This is the time in which we can dream of what may be. Many of us will not see the other shore of the *Long Dark*. But some may. As Duane Elgin writes, "Now I see myself planting seeds of possibility, but without expectation that I will live to see them blossom in a new summertime or partake of their fruits in the harvest of a distant autumn. My approach now is to trust the wisdom of the Earth and the human family in bringing forth another season of life." That is an elder's blessing. We live for what may be, knowing we may never see the fruits.

The only way out is through and the only way through is together. This is a collective initiation. This is the gestation period for a possible planetary community. We are the midwives, the elders, the guides to our future life. It is a good time to be alive.

—Francis Weller
Russian River Watershed
Shasta Bioregion

PART I

Our World in Great Transition

We do not inherit the Earth from our ancestors,
we borrow it from our children.

—Native American Wisdom

Humanity's Initiation and Transformation

We often forget that we are nature. Nature is not something separate from us. So, when we say that we have lost our connection to nature, we've lost our connection to ourselves.
—**Andy Goldsworthy**

If you have read the powerful preface by my good friend Francis Weller, then you know the people of the Earth have entered a time of great transition—a time of collective initiation as we move through great sorrows to awaken new potentials. We are moving through a painful birth as a species as we grow into our collective maturity. *Choosing Earth* is intended for mature and resilient humans who are ready to reach deep and explore our world in this unprecedented transition.

Looking ahead, I see two certainties: First, the future is profoundly uncertain because so much depends on the choices we now make, individually and collectively. Second, the world of the past is gone. We cannot return to the "old normal" because it was never "normal"—it was abnormal with extremes of overconsumption, species extinction, melting ice-caps, dying oceans, severe droughts, massive wildfires, profound alienation, extreme inequities, and much more. Great loss and great transition are now upon us. There is no going back. There are no do-overs. We cannot re-freeze the polar ice-caps and recreate the congenial climate of the past ten-thousand years. We cannot refill ancient aquifers that have been pumped dry. We cannot swiftly restore the complex ecology of the past and bring thousands of species of animals and plants back to life. We cannot stop the momentum of sea-level rise even if we stop CO_2 emissions now. We cannot undo the overshoot created by overconsumption and depletion of the Earth's resources. A profound initiation is underway that will shake and transform us to the core. Great promise and possibility call us, drawing our vision beyond tragedies of our own making.

We are creating this rite of passage. This is not the time to falter and hang back. We are challenged to step up together and move forward with courage as if our lives depend on it. They do. Still, many hesitate. We may think we have more time—assuming the pace of change in the past is an accurate measure of the rate of change in years ahead. This is not the case. The rate of change is accelerating as powerful trends reinforce one another and converge in an immense wave of change that is washing away the world of the past. No longer can we safely extrapolate the rate of change we experienced in the past as a measure for the future. We are out of time. Our very existence depends on looking through fresh eyes at our world in profound transition.

We may also hesitate because we think new technologies will save us from the discomfort of making fundamental changes in our lives. However, the forces of change are so deep and powerful that we require all of our technological ingenuity and *much more*. Technology alone will not save us. The many challenges we face call for a deep shift in the way we relate to all of life: these include the food we eat, the transportation we use, our levels and patterns of consumption, the work we do, the dwellings in which we live, the education we acquire, the way we treat people of different races, genders, cultural, and sexual orientations. We are called to reconfigure our lives both individually and collectively. The magnitude of change required by our times is nearly inconceivable. The editors of the respected journal *New Scientist* offered this assessment of the work that lies ahead:

> "It will arguably be the largest project that humanity has ever undertaken—comparable to the two world wars, the Apollo program [to put a human on the moon], the cold war [with a nuclear arms race], the abolition of slavery [which included a civil war], the Manhattan project, the building of the railways and the rollout of sanitation and electrification, all in one. In other words, it will require us to strain every muscle of human ingenuity in the hope of a better future, if not for ourselves then at least for our descendants."[5]

But how can this happen? What is a realistic path forward to accomplish this magnitude of change? That is the journey explored in this book.

Still, people ask me: Why look ahead? Why think about a future of gloom and doom? Can't the future take care of itself? Why not be happy, be kind, and live in the now? We cannot predict what is going to happen. Life has so many surprises, how can we envision what lies ahead? Doesn't imagining the future take us away from living in the here and now? We are small beings who cannot change what is happening, so why bother with what we cannot change?

Why should we look ahead? What can be gained? Here's why: We now live in a tightly interdependent and transparent world where our individual fates are tied directly to the fate of the planet. Given this reality, I encourage us to look ahead and, with freedom and creativity, to consciously choose our future in order:

> 1. to prevent the **functional extinction** of humanity and much of the rest of life on Earth;
>
> 2. to avoid imprisonment in the endless darkness of an **authoritarian** world;
>
> 3. to grow up and move, with maturity and freedom, into a **transforming** world.

To choose to not look ahead is a profound choice. "Let the future take care of itself" is the mindset of an adolescent stage of life. Our world is calling us to grow up and to take charge of moving into our early adulthood and to care for the wellbeing of all life. The future is not impenetrable—it is graspable and malleable in our minds and intuition. If we see it, we can choose it. If we don't look ahead, we are unprepared. Unprepared, we respond superficially. Acting without depth, we are overwhelmed by avalanches of deep change.

I understand how looking into the depths of change just ahead confronts our psyche and soul. Our times are not for the faint-hearted. This is not the time to live small and retreat from the world. These are times for living into the immensity of being as citizens of the living cosmos and for consciously choosing our future for living on the Earth.

Stepping back for perspective: I began exploring deeply the challenges ahead a half-century ago in 1972, when working as a senior staff member of the Presidential Commission on Population Growth and the American Future.[6] Our mandate was to look thirty years ahead and consider how and where growing numbers of people might live. At that same time, the seminal book *Limits to Growth* was published and our commission began exploring the closing circle of the world's ecology. Work on the presidential commission revealed not only the limits to growth of our nation's consumer economy, but also the limits of our government's ability to even think about making a transition to a sustainable future.

After the commission finished its project, I began working for the "futures group" of the Stanford Research Institute (now SRI) think-tank. A personal story further illustrates the unresponsiveness of government bureaucracies to major threats to our future. I first learned about global warming as an existential threat to humanity in 1976 while working as a senior social scientist on a year-long project for the National Science Foundation at SRI International.[7] I was part of a small team looking for unexpected, future challenges that could wipe us out from the blind side. In support of this project, I attended a briefing on climate change at the Department of Energy in Washington, D.C. At the briefing, we were told that, if present trends continued with CO_2 build up, in another 40 to 50 years it would create serious problems of global warming for the planet. Despite this somber warning, energy officials discouraged us from including global warming in our report. They reasoned that this would not grow into a crisis for nearly fifty years and this would give the political process plenty of lead time to mount a response. Not only did we not include global warming in our report, the government officials in charge of our work decided the report was too controversial for public consumption and was put away from easy access by politicians and the public.

Now, nearly a half-century later, we can see the results of decades of delay: As we anticipated, the world is under assault with a dramatically changing climate and the faltering governance of civilizations. Given this experience, I do not expect existing institutions—government, business, media, and education—to rise quickly to the unprecedented challenges we now face. As I wrote in another report to the president's science advisor, our large, highly complex bureaucracies are not configured to respond with the speed and creativity necessary to meet the challenges of our perilous times.[8] For that reason, I place my greatest faith in the people of the Earth organizing ourselves from the local to global level and, together, swiftly learning and choosing our way into a sustainable and purposeful future.

Based on these kinds of experiences, I left the futures group at SRI in 1977 and turned to writing a book on the theme of *Voluntary Simplicity*. I began with a half-year of solo meditation with the intention of bringing together all that I had learned—both inner and outer aspects of my life—and returning to the world as a whole person. Intensive meditation brought new insight into humanity's future and the understanding that the decade of the 2020s would be the time when humanity would be forced to make a pivotal turn in our evolution as a species.[9] Based on this understanding, since 1978 I have been writing and speaking about the decade of the 2020s as the pivotal time when humanity would be confronted with making a turn and choosing a new pathway into the future. Now, this fateful decade has arrived.

Absorbing the scope, speed, and depth of change of our world in unprecedented transition has been deeply confronting. Sorrow has been my faithful companion—anguish my teacher. I've been humbled by the intensity and immensity of suffering growing in the world, knowing this tsunami of sorrow will break our hearts and, at the same time, open us to our higher humanity. Although writing has been a major part of my life journey, this has been a soulful challenge beyond the reach of words. My writer's desk has become an altar to despair as I recognize and accept all that will perish as humanity moves through this great transition.

As I've stepped back to look, again and again, seeking perspective for what is unfolding, I know that I am writing this book from a privileged perspective—of a white, male member of a highly industrialized Western culture and nation. Although my roots are in a small farming community in Idaho, I have lived most of my adult life in a modern, urban-industrial setting. Yet, as I try to find my place in our world in profound transition, I find myself returning to my roots as a farmer. Now I see myself planting seeds of possibility, but without expectation that I will live to see them blossom in a new summertime or partake of their fruits in the harvest of a distant autumn. My approach now is to trust the wisdom of the Earth and the human family in bringing forth another season of life.

We are creating a rite of passage for ourselves as a species—but what kind of passage and to where? Could the enormity of our imagined loss be a catalyst for unimagined gain? Could a new human alloy—rich with aliveness and potential—emerge from the furnace of the superheated decades we have now entered? These questions are at the heart of this book.

From a state of trust, *Choosing Earth* explores the collapse and transformation of the world we have constructed over the past ten-thousand years. Acknowledging the breakdown and fall of our world is the first step on our passage to a new life. It is vital we not turn away from engagement with collapse, but embrace this reality as integral to our initiation into adulthood as a species. The grief and sorrow we experience are awakening us to deep transformation. We are asked to leave the past behind as the world is already unraveling—fraying and coming apart—and we must make preparations for free-fall and collapse. In the words of Marianne Williamson, "Something very beautiful happens to people when their world has fallen apart: a humility, a nobility, a higher intelligence emerges at just the point when our knees hit the floor."

Humanity's rite of passage will lead us into a new understanding of the reality we inhabit, the nature of ourselves as beings of both earthly and cosmic dimension, and the extraordinary evolutionary journey on which we have now embarked. *Choosing Earth* is choosing life. The breakdown and collapse of our world contain the terrifying reality that our species could so devastate the biosphere that we will become functionally extinct. The breakdown also contains the potential to move through a time of great initiation and into a new era of possibility. Together, we could choose a path that serves the well-being of all life. Together, we could move through great loss, grief, and sorrow and allow our knees to hit the floor and then, with humility, rise on a path of great transition.

It is vital to recognize where we are on our evolutionary journey. We have reached a critical threshold where we cannot go back and must go forward. To simply adapt to where we are is to surrender to evolutionary stagnation and our functional demise as a species. If we don't choose to go through these difficult times and grow into our collective maturity, we will leave a legacy of ruin for the Earth and ensure our functional extinction as a species. Do or die. *We have no future without our maturity.* If we move *through* our adolescence to early adulthood, we can discover untapped uplifting potentials. Alternatively, we can abandon our evolutionary advance by holding onto a shallow and shrunken view of humanity and our journey. Are we comfortable with the prospect that our legacy as a species may be a few decades of consumer comfort for a fortunate few? Are we at ease with the view that *Homo sapiens* represents a thread of life that faltered and failed because we were so preoccupied with materialistic pursuits that we did not blossom into maturity? We know we are better than this, so don't lose heart!

We cannot reach the heights unless we also recognize the great depths. When all seems lost—when there is nothing left to lose—we can let go of the past and rise to new heights and potentials. Now is a time of great choice for our world. We are called to greatness as a species—to realize our collective maturity as a planetary community. Nothing will ever be the same. Transformed by sorrows, we can move into a new world. A fresh understanding of human identity and our evolutionary journey calls us forward, drawing us into a future of immense possibility. A path of evolutionary uplift is both a gift and a choice. Our time of collective choice has profound consequences that will carry forward for thousands of years. There is no way around our rite of passage—*there is only going through*. We created these times and we can move through them, consciously, creatively, and courageously. The journey ahead is so pivotal that it's worth fully investing our lives in a transforming outcome. The odds are long, but the rewards are great.

Growing Resilience in a Transforming World

We now confront such enormous challenges that we can quickly feel overwhelmed. We can empower ourselves by exploring meaningful actions to take in our everyday lives.

1. Choose aliveness—Choose activities that bring you alive: walking in nature, dancing, playing, making music, nurturing relationships, making art, and connecting with animals. Create an altar of gratitude. Offer affirmations and prayers for plants, animals, places, and people. Become a role model of gratitude and aliveness for younger people.

2. Cultivate your "true gifts" — We each have "near gifts" and "true gifts."[10] Near gifts are things we are relatively skilled at doing. We often make a living with our near gifts. True gifts express our natural talents and skills — activities where we naturally shine. Developing your true gifts is an exercise in becoming more fully alive and connected with the world.

3. Develop your consciousness — The quality of your awareness is enormously important for navigating our changing world. Cultivate an awakening heart-mind through practices such as meditation, yoga, prayer, dialogue, or other mindful activities. Become an ever-more conscious participant in life.

4. Be locally informed — Get to know your local ecosystem. Learn the trees, flowers, birds, and other animals that abound locally. Recognize foods grown locally. Explore and experience nature when you take a walk. Find ways to support your local ecosystems and healthy local farms and businesses.

5. Protect and restore nature — Take small actions to help restore nature and the miracles of life. Be curious and learn how you can protect the natural world around you. Because nature cannot advocate for itself, become a voice for wild plants, trees, and animals and their preservation and restoration.

6. Grieve losses — Create an altar in your home, with images and objects, to acknowledge what we are losing (trees, flowers, animals, seasons, places, etc.). Organize a simple grieving ritual with others and have everyone share what they are mourning (what is lost or forgotten) — speak deeply, sing songs, read poetry, and share art.

7. Practice reconciliation — Acknowledge your advantages, explore what this means with a group of trusted friends or peers. Bring curiosity and compassion to divisions of gender, race, wealth, religion, and sexual orientation.

8. Choose simplicity — Buy fewer things, give more away, eat lower on the food chain, travel less by plane, reduce or change your commute, and share your resources with others in need.

Cultivate meaningful friendships, share simple meals, take walks in nature, make music, do art, learn to dance, develop your inner life.

9. Organize a study group—Step back and look at our world in a time of unprecedented transition. Use this book and study materials on the *Choosing Earth* website www.ChoosingEarth.org to explore with others. Avoid jumping into problem-solving or blaming and leave plenty of room for feelings to be expressed. Explore ways to embody this knowing.

10. Support others—Encourage and assist individuals and communities directly impacted by climate change, racism, species extinction, inequalities, resource depletion. Make your life a statement of care by acting to protect the local ecology. Volunteer for service organizations—a local foodbank, homeless shelter, or regenerative gardening or farming.

11. Cultivate communication—Become a voice for the Earth and humanity's future. Contribute to newsletters, blogs, articles, videos, podcasts, and radio to bring your voice and views to our endangered future. Help awaken our social imagination to the choices we have for maturation, reconciliation, community, and simplicity.

12. Become a compassionate activist—Join with others working for deep transformation. Search the internet to find organizations that fit your interests. Whether local or global, find a community that supports you in bringing your true gifts into the world at this critical time. Give of your time, your love, your talents, and your resources.

13. Hold institutions accountable—hold major institutions (business, media, government, and education) publicly accountable for recognizing and responding to the critical challenges facing the Earth and humanity's future. Accountability can be challenging because we all are embedded within these institutions—which means we hold ourselves accountable, as well.

Seemingly small actions in our personal lives provide grounding for ourselves and a shining example to others.

Never doubt that a small group of thoughtful, committed citizens can change the world; indeed, it's the only thing that ever has.
—**Margaret Mead**

Both visionary optimism and unflinching realism are important. Global surveys show that most people recognize to some extent the great dangers and difficulties that lie ahead. A 2021 survey explored the views of ten-thousand young people, from 16 to 25, in ten countries around the world and found deep anxiety about the future.[11] Three-quarters said they thought the future was frightening and over half (56 percent) said they think humanity is doomed! Two-thirds reported feeling sad, afraid, and anxious. Nearly two-thirds said governments are betraying and failing young people. Most think humanity has failed to care for the planet (83 percent). This is a stunning assessment of our condition. Young people around the world are losing confidence and trust in the world being left to them. A profound break with human history is already present for youth who no longer feel at home in our changing world.

Another global survey in 2021 questioned more than a million people in fifty countries. The *Peoples' Climate Vote* was the largest survey of public opinion on climate change ever conducted. Overall, this massive survey found that 59 percent said there is a climate emergency, and the world should "do everything necessary" to meet this global crisis.[12] A deep recognition now exists that the fate of the Earth hangs in the balance.

Although we face a profound climate emergency, the challenges we confront reach far beyond climate—the entire web of life is under assault. A mass extinction is underway, impacting animal and plant life on land and in the world's oceans. Agricultural productivity is falling at the same time that human populations are growing, and this disparity is producing widespread food shortages. In turn, famines force mass migrations of people toward more resource-favorable places. Overwhelming numbers of climate refugees lead to civic breakdowns, as countries and governments are unable to cope. Plants and animals are going extinct, unable to keep up with shifting climate and ecosystems. The Amazon rainforests are being transformed into diminished ecosystems of scrub and brush.

Roughly half of the people on Earth live on the equivalent of two dollars a day or less. The suffering unleashed by this time of great transition disproportionately impacts poor people, indigenous people, and people of color. Extreme inequities in wealth and well-being trigger growing conflicts, as the dispossessed try to climb out of deep poverty. Beyond a climate crisis, we are in a whole-systems crisis for the Earth. The entire fabric of life is being torn apart and profoundly wounded.

Numerous times, the Earth community has been warned about these critical trends. The most vivid and stark warning was delivered decades ago. In 1992, more than 1,600 of the world's senior scientists, including a majority of living Nobel prize-winners in the sciences, signed an unprecedented document titled *Warning to Humanity*.[13] In their historic statement, they declared that "human beings and the natural world are on a collision course . . . that may so alter the living world that it will be unable to sustain life in the manner that we know." This is their warning:

> "We, the undersigned senior members of the world's scientific community, hereby warn all humanity of what lies ahead. *A great change in our stewardship of the earth and the life on it is required if vast human misery is to be avoided and our global home on this planet is not to be irretrievably mutilated.*"[14] [emphasis added]

In re-reading this conclusion, my thoughts return to a few key words in their warning where scientists state that, if great changes are not made in our stewardship of the Earth, the planet will be "*irretrievably mutilated.*" Those last two words reverberate in my being. What does "irretrievably mutilated" mean for countless generations ahead? The Earth forever disfigured, permanently damaged, maimed, and mutilated for all time? Will failure for responsible planning and stewardship be our legacy to future generations?

More than thirty years have passed since this blunt warning was issued. Our response to this dire threat facing humanity has been painfully slow and can be summarized in four words: *Too little, too late*. We have allowed critical trends to race ahead, leaving us behind. The pace of breakdown is far faster than the pace of our collective response for repair and healing. We are out of step with reality. The Earth's ecology has been unraveling for more than a half-century and gradual breakdown is now cascading toward collapse. We are being overtaken and overwhelmed. We must prepare for collapse as well as for evolutionary advance.

We are challenged to wake up together and respond with maturity to a world in great transition. It is not just the speed of change that overwhelms, it is also the scale and complexity of change. We confront a multitude of accelerating crises—growing climate disruption, spreading water scarcity, declining agricultural productivity, increasing inequality in wealth and well-being, rising numbers of climate refugees, widespread extinction of plant and animal species, dying oceans polluted with plastics, and expanding bureaucracies of overwhelming size and complexity. Our world is spinning out of our control. New ways of living and being on the Earth are critical.

Collapse is inevitable.
Moving through collapse is a choice.

We humans have already gone too far, and momentum is too great, to avoid breakdown and collapse. We are already in profound overshoot—stealing from future generations and disrupting the well-being of all life. We can continue this way for only a short while longer. If we continue robbing from the future, the collapse of human systems and eco-systems is our inescapable destiny. However, if we collectively witness the world of devastation that is growing exponentially, we can choose together a more favorable future for all life. The alternative is devastating ruin and the functional extinction of humans on Earth.

Stepping up to this level of change is completely unprecedented and will require nothing short of a revolution in collective effort by humanity. However, even this stunning description fails to reveal the depth of practical change that is essential. We need a sweeping transformation of energy production and use to avoid disastrous global warming. Scientists estimated that the human community had to halt increases in fossil fuel emissions in 2020, and then cut them in half by 2030, and then cut them in half again by 2040, and then come to a net zero of carbon emissions by 2050.[15] The *entire world* must either eliminate or offset carbon pollution by mid-century. This means that:

- By 2050, no home, business, or industry will be heated by gas or oil or, if they are, their carbon pollution must be offset.
- No vehicles can be powered by diesel or gasoline.
- All coal- and gas-power plants must be shuttered.
- Even if the world succeeds in generating all its electricity from zero-emission sources, such as renewable energy or nuclear power, electricity makes up less than one-third of current fossil fuel consumption. Therefore, other energy-intensive users of fossil fuels—particularly those used to manufacture steel and concrete—must be fueled by renewable sources.

While a complete rebuilding of the entire energy infrastructure of the world within a few decades is vital for a workable future, it is far from sufficient. In addition, a deep and profound transformation is required in virtually every aspect of life—the food we eat, the skills we develop, and the work we do, the homes and communities in which we live, the media messages we produce and receive, the local-to-global conversations we develop, the values of economic fairness and social justice we share, the leadership offered across diverse institutions (political, religious, media, non-profit), and more. *Building an entirely reconfigured society, economy, culture, and consciousness is our only pathway for avoiding the irretrievable mutilation of the Earth.*

How can we implement a massive and complex transformation of lifestyles to bring us into balance with nature's limits? Currently, people in the wealthier countries and regions of the Earth consume far more than their fair share of the planet's resources. This overconsumption deprives a majority from their fair share and condemns them to poverty and a disproportionate level of climate-induced suffering. This inequity is so extremely discriminatory and unbalanced that it cannot endure. It will be immensely challenging for those with high-consumption lifestyles to deliberately limit their drawdown of resources and to share their wealth with those less economically privileged. Humanity's survival requires a lifestyle revolution where the wealthy choose ways of life that are materially far more restrained in the use of the Earth's limited resources and far more generous in fostering the well-being of those who are materially impoverished.

A transformational shift in ways of living is more than a matter of moral justice and fairness—it is also essential for preventing all-out class warfare over resources. If we are going to work together as a human community, then those accustomed to being in positions of authority and power (as a result of class, gender, race, geography, age, ability, education, etc.) must step forward to lift up the lives and voices of the global majority (poor folks, indigenous communities, and other long-suffering and oppressed groups). Only then will it be possible to create meaningful, systems-level changes, including the redistribution of resources that will liberate the global majority from being forced by survival pressures to focus only on their urgent, short-term needs.

In addition to great concern for the *magnitude* of change, alarm is growing with regard to the *speed* of change, particularly with respect to climate disruption. In the past, scientists thought it would take centuries, if not thousands of years, for climate to swing into a different configuration. It came as a profound shock to discover that a major shift can occur within "a matter of decades or even less."[16] To illustrate, a period of global cooling, called the Younger Dryas, occurred roughly 11,800 years ago (likely the result of an asteroid breaking up in the atmosphere) and was followed by a period of abrupt warming, estimated to be roughly 10°C within a matter of years![17] Although such astonishingly rapid levels of temperature change are not currently predicted, this example does reveal our vulnerability if we ignore historical variations. Current institutions of government and policy thinking would be completely unable to step up to deal with such abrupt climate change. Most governing institutions are designed to perpetuate the past, not to move swiftly into a transforming future.[18]

In addition to the magnitude and speed of change, we are obliged also to recognize the *depth* of change required for this time of great transition. "Choosing Earth" means choosing a new relationship with the Earth—which means choosing a new relationship with the totality of life. By our own hand, we have created the conditions that force us to look at our behaviors more consciously and to deliberately choose our path ahead, both as individuals and as an entire species. The breakdown of life on Earth brings with it the breakdown of our collective psyche. *Eco collapse brings with it ego collapse.* Fundamental advances in our collective psyche are now imperative. We cannot repair the Earth without healing ourselves and our relationship with the rest of life. Gus Speth, former director of the Council on Environmental Quality, has described the nature of our challenge clearly:

> "I used to think the top environmental problems were biodiversity loss, eco-systems collapse, and climate change. But I was wrong. The top environmental problems are selfishness, greed, and apathy . . . and to deal with those we need a spiritual and cultural transformation—and we scientists don't know how to do that."[19]

Although politicians and the mass media present what is happening as an ecological crisis, it goes far deeper than that. Not only are we running into an "ecological wall" or the physical limits of the Earth's ability to sustain humanity, we are also running into an "evolutionary wall" where we confront ourselves and the consciousness and behaviors that lead us into overshoot and collapse. An evolutionary wall presents humanity with an identity crisis: Who are we as a species? What evolutionary journey are we on? Do we have the inner potentials to meet the demands of the outer world? Can we rise in our maturity and grow into a healing and healthy relationship with the Earth?

If we do not step up to meet both the outer and inner challenges of our times, we seem destined to follow the example of more than twenty great civilizations that have collapsed throughout history—including Roman, Egyptian, Vedic, Tibetan, Minoan, Classical Greek, Olmec, Mayan, Aztec, and a number of others. Our vulnerability is made starkly clear as we acknowledge the breakdown and disintegration of these great civilizations of the past. However, the current situation is unique in one key respect—human civilization has reached a global scale and encircles the Earth as an interdependent system. *The circle has closed. Now the simultaneous downfall of all the intertwined civilizations on Earth is threatened.* Nothing in humanity's history prepares us for a rapid collapse of the tightly interconnected civilizations throughout the world.

An extraordinary push and unprecedented pull are at work in these transitional times. If we look only at the push and ignore the pull, it places our journey in great peril. To visualize this process, imagine pushing on a length of string. Pushing ahead, the string will bunch up in front of us and create a tangle of knots. Then imagine simultaneously pulling on the string—it no longer bunches up in a jumble, but can move forward in a line of progression. In the same way, if we understand and respect both the pushes and pulls of our times, we can move ahead without getting completely entangled in the process.

If we consider only the unyielding push of the climate crisis combined with other adversity trends, then our efforts will produce complex knots and we can easily become mired in confusion and despair. However, if we deepen our vision to include the pull of opportunity, then we see the possibility that we can move ahead with stunning speed. The pull of opportunity does not eliminate the enormous challenges we face. Instead, by recognizing and working with both the powerful push of necessity and the remarkable pull of opportunity we can find the courage, compassion, and creativity to work through the difficulties of transition.

To see our time of great transition more clearly, let's take a whole-systems view with three perspectives:

- **Look Wide**: Look broadly, beyond single factors, and consider a wide range of trends as an integrated system—climate disrupting, population growing, refugees migrating, resources depleting, species dying, inequities increasing, and much more. Looking wide provides us with a much clearer picture of change often missing when attention is focused on only a single area.

- **Look Deep**: Look into the depths beneath the outer world—include the inner dimensions of change, such as our evolving psychology, values, culture, consciousness, and paradigms. The outer world reflects our inner condition. By evolving our inner world, we simultaneously evolve our capacity to evolve the outer world.

- **Look Long**: Look far into the future—much farther than the short run of the next five or ten years. Trends that are uncertain and ambiguous in the short run become much clearer when extrapolated to the longer run where their impact is much more distinct and well-defined.

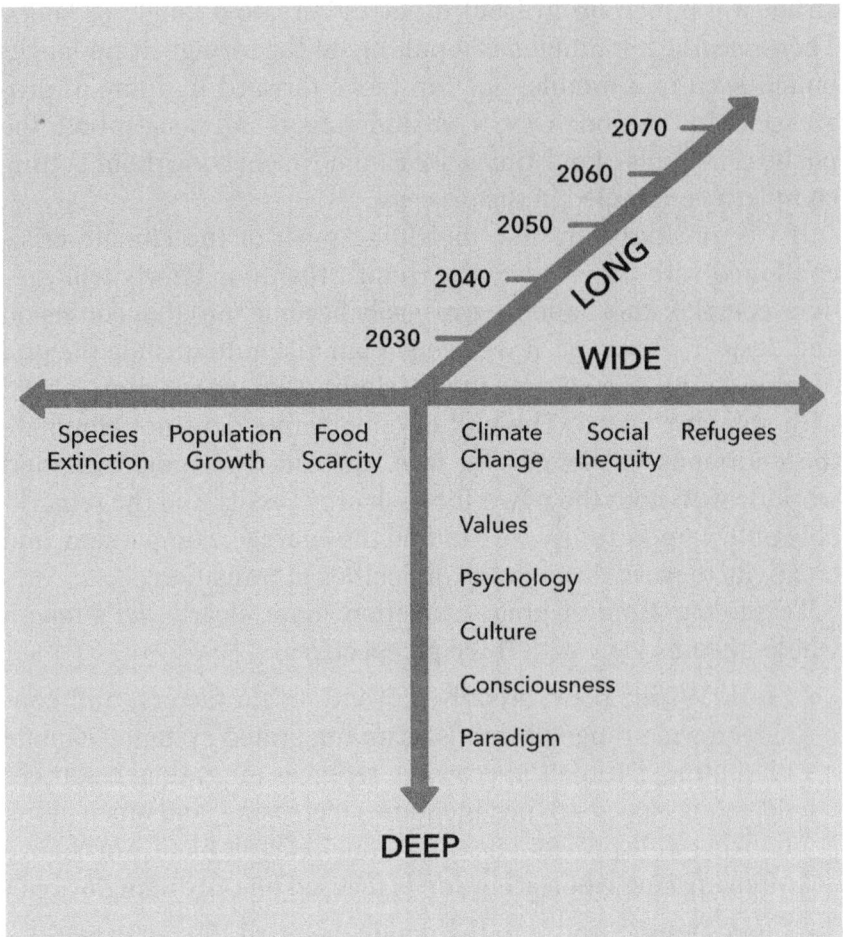

Figure 1: Wide, Deep, and Long

When we look wide, deep, and long, we see more clearly the pivotal time in history we have entered and how we can move more deliberately beyond our time of great transition. From this whole-systems perspective, we see we can either find uplift and rise to a new way of living or we face downdraft into collapse and ruin. Stark choices are now before us. Not decades or centuries ahead, but *now*. We have run out of time.

PART II

Three Futures for Humanity

"Forces beyond your control can take away everything you possess except one thing, your freedom to choose how you will respond to the situation."

—Victor Frankl

"We are pilgrims together, wending through unknown country, home."

—Father Giovanni, 1513

Extinction, Authoritarianism, Transformation

It is important to acknowledge how open and vulnerable our future is at this singular moment. We have entered an extraordinarily rare interval in history—a choice-point in our collective journey—a space between the past and the future where the lives of (hopefully) countless generations ahead will be profoundly impacted by choices we make now. We cannot predict where humanity will go from here, for a simple reason: Our future depends on our conscious choices—or failure to choose—both individually and collectively. Our evolutionary journey will either become conscious of itself or descend into darkness. We are at a turning point in history—a time that will forever be remembered, as we either rise in our maturity as a species-civilization aware of our responsibilities or we descend into ruin and obscurity.

A crisis calling us to take urgent action did not have to be our destiny. Nearly half a century ago, in the 1970s, humankind squandered an opportunity to gradually adapt to a radically changing future. That's when the immense challenges we face today were first recognized. At great cost, we consumed the margin of extra time in order to keep the status quo alive for a few additional decades.[20] Now it is too late to choose a path of gradual change.

Having used up our breathing room for gradual adaptation, humanity now confronts profound consequences if we don't respond swiftly and make sweeping changes in how we live on the planet. Within a few decades, large portions of our world will no longer be fit for human habitation. Extremes of drought, floods, and storms will become common. Famine and disease will shake humanity to the core. Hundreds of millions of climate refugees will be on the move, looking for places to live. The mass extinction of animals and plants will forever impoverish the ecology of the Earth. Options for the future are becoming severely limited. The time for gradualism has ended.

Below, I explore three major pathways that represent our clearest options for the future. It is important to recognize that *all three of these pathways begin with the same underlying trends and conditions—a dynamic process called "collapse."* Because I give considerable attention to "breakdown" and "collapse," I want to clarify the meaning of each. These terms are often used interchangeably, but can be understood quite differently:

- **Breakdown** means that linkages in key systems are coming apart and failing. Supply chains for delivery stop operating for significant periods. Power outages occur. The water stops running at times, and its purity is doubtful. Fire and police departments close periodically because they cannot pay people. Breakdown refers to the disintegration of whole systems into their component parts that, while disrupting and damaging to health, employment, and access to essential services, also creates opportunities for new living configurations. By disrupting business-as-usual, breakdowns create openings for rebuilding in new ways that can be healthier and more resilient. Breakdowns can be catalysts for creativity and spur innovation—for example, in rebuilding and retrofitting communities with local economies that support more resilient approaches to living.

- **Collapse** is far more serious than breakdown because it describes a process of ruinous downfall of communities, cities, and civilizations. With collapse, society fails completely—as housing, transportation systems, water and sewage systems, and more, fall into jumbled chaos. Collapse is the catastrophic failure of the system *and* its components. Collapse leaves both (the system and its components) in a condition of rubble—a junkyard of broken systems of all kinds—transportation, communication, and civic services. Collapse produces a very difficult foundation (physical, economic, psychological, social, and spiritual) from which to build a promising future of inclusive, sustainable well-being.

Here, I offer two graphic descriptions of what collapse could mean for the world. First is Venezuela. Once one of the economic miracles of South America, with one of the largest reserves of oil in the world, its economy has collapsed in the last few years with devastating consequences:

> "Desperate oil workers and criminals are stripping the oil company of vital equipment (vehicles, pumps and copper wiring), carrying off whatever they can to make money.... Venezuela is

on its knees economically, buckled by hyperinflation and a history of mismanagement. Widespread hunger, political strife, devastating shortages of medicine, and an exodus of well over a million people in recent years have turned this country, once the economic envy of many of its neighbors, into a crisis that is spilling over international borders."[21]

Second, here is a description of collapse in Haiti, where gangs rule much of the country:

"With more than a third of Haiti's population of 11 million already in need of food assistance, rampant criminal gangs have paralyzed fuel deliveries, without which economic activity—and the availability of food and medical care—has ground to a halt. The government is an empty shell and often in league with the gangs who have seized control of entire neighborhoods and critical roadways. An epidemic of kidnappings has spread unchecked. Mayhem is enveloping nearly every aspect of daily life. Massacres, gang rapes, and violent arson attacks on neighborhoods are widely reported."[22]

With *breakdowns*, the components of life remain sufficiently intact to be re-assembled into new configurations that can work—potentially, even better than before. However, *collapse* requires building a new operating system on the scrapheap of ruined infrastructure, shattered institutions, and a devastated ecology.

The aftermath of devastating wars illustrates the ability to recover after a systemic collapse—*if* a functioning ecosystem remains intact. As a prime example, we need only look at the era after World War II when nations rebuilt from the rubble and ruin of war. Germany suffered massive devastation and the collapse of its economy, society, and infrastructure. Yet, the post-war era was followed by swift rebuilding. As this illustrates, the term "collapse" describes a condition of nearly complete wreckage of a country, economy, and society, but this does not mean a final ending. What emerges from the dynamic process of collapse depends greatly on the ability of people to mobilize themselves swiftly and constructively. In a similar way, which future pathway will eventually emerge from a planetary-scale collapse depends greatly on the extent to which citizens of the Earth are able to mobilize themselves with rapid and creative responses to build a new future.

I imagine that, after planetary collapse and the breakup of nations, power will be widely dispersed among a bewildering conglomeration of groups and communities, each mobilizing themselves for survival. A patchwork of communities and competencies will likely emerge with no one in overall control. Some could have more fighting power with access to powerful weapons and others could have more economic power with access to important resources and skilled people. Some communities could be self-organizing and self-governing, while others could be managed by "overlords" and their armies. The overall condition could be one of continuous bargaining, trading, fighting, and compromising. Fragmentation seems likely to be so great that no one would be able to get the upper hand and exert overall control. The struggle for power in a world that requires diverse skill sets creates a crucible for discovering new ways of living. Breakdown and collapse could produce settings for intense experimentation. A new human "alloy" could emerge from the heated competition among communities and provide the foundation for building larger, regenerative societies.

The dynamic nature of "collapse" reveals a key question: *Will the people of the Earth be willing to truly step up and halt the breakdown of the biosphere before the planet becomes completely uninhabitable?* To set the stage for a deeper inquiry, here are brief summaries of how collapse could develop in three different futures:

- **Functional Extinction** could be the product of unrestrained global warming, producing an uninhabitable climate and the mass extinction of most forms of life, combined with the collapse of civilizations due to famine, disease, and conflicts. The devastation of the Earth's ecosystem together with the destructive downfall of civilizations could push humanity to the very margins of existence. Humanity could become "functionally extinct" while continuing to live at the edges of survival—but so diminished in numbers and capacities that we would fall beneath the threshold of evolutionary significance. Admittedly, humanity could move even beyond functional extinction to *actual* extinction if we alter the Earth's climate beyond what biology can tolerate. In short, we could cook ourselves to death and become completely extinct.

- **Authoritarianism** could emerge as a sweeping alternative if humanity were to pull back during the early stages of planetary collapse and accept highly intrusive forms of constraint. Artificial intelligence could empower sophisticated forms of monitoring and control that reduce the severity of collapse with extreme limits placed on social interactions. Regimented forms of civilization could become dominant with the lives of citizens highly controlled by some powerful authority. Because authority is concentrated, the masses would end up being at the mercy of a few.

- **Transformation** could emerge if people were prepared to adapt swiftly and orient toward a more sustainable, inclusive, and compassionate future, with a high level of collective maturity and collaborative living. With anticipation and imagination, the more extreme expressions of collapse could be moderated, and our maturity awakened to support diverse expressions of uplift for constructing a purposeful and regenerative future.

Three key insights emerge. First, *all three of these pathways begin with breakdown and collapse.* The difference is not in the driving trends that lead to initial collapse, but in how we mobilize ourselves in response to those unfolding trends. Second, "collapse" is not a singular condition, but a dynamic process from which recovery can emerge. So far, the Earth has undergone five mass extinctions and life has recovered each time, generally over a period of millions of years. Humanity's ruin of the Earth does not mean that all life would be ended, but it could very well mean a recovery time measured in tens of thousands or even millions of years—which, in turn, means that humanity would likely become extinct, just as did the dinosaurs and many other forms of life in a prior mass extinction. Third, all three pathways will be present to varying degrees in the coming decades of turbulent transition—which leads to a pivotal question: *which of these three scenarios will predominantly guide us into the far future?* With this introduction, let's briefly explore each of these pathways ahead.

Future I: Extinction

The world has to wake up to the imminent peril we face as a species.
—Inger Andersen, director of the U.N. Environment Program

In this pathway, the world continues along its business-as-usual pathway, largely in denial of the great dangers rapidly developing and reinforcing one another, producing a severe, whole-systems crisis. Much of the materially developed world remains absorbed in a collective trance of consumerism, accepting the view we are separate from one another, from nature, and the universe. Although diverse movements for transforming society and restoring ecology might emerge, they are too small and too weak to penetrate the distraction and denial of the majority. As a result, we fail to recognize the looming perils and move toward collapse and functional extinction. To repeat, "collapse" is not singular condition, but a dynamic process that grows with increasing severity. Here is how I envision a spectrum of collapse across five stages, moving from initial breakdowns to full extinction:

1. Widespread Breakdowns. Diverse systems unravel and come apart. Supply chains for goods and services break down. Essential services such as police and fire protection, sanitation, education, and healthcare become increasingly unreliable. Climate continues to warm, species die off, mass migrations occur, and water shortages become critical. Breakdowns can serve as a catalyst for creative adaptation, so this stage still has great potential for pulling back and developing more viable approaches to living on the Earth.

2. Collapse Underway. Supply chains and vital systems break down throughout the world. Ecosystems fail, oceans no longer support life, agricultural productivity drops, starvation and migration increase. The potential for regeneration of human systems and ecosystems still exists, but becomes increasingly costly and unaffordable. Although this scenario involves a deep wounding of Earth's and humanity's future, nevertheless we can still recover from these destructive times.

3. Full Collapse. The hard collapse of human systems combines with irretrievable damage to the biosphere. It is impossible to regenerate the ecosystems of the past; instead, we are forced to rebuild from a profoundly injured ecological and human foundation in an attempt to create a healthy biosphere from what remains.

4. Functional Extinction. Humans are no longer a viable species. Sperm counts drop to near zero and we are unable to reproduce ourselves as a species. Relentless pandemics proliferate unchecked, further eroding humanity's foothold for survival. Global warming renders the Earth inhospitable and largely uninhabitable. The overall ecosystem is devastated and mutilated beyond recognition. Pockets of humanity remain, but a significant human presence has vanished, leaving only a few survivors behind locked in a struggle for survival amidst the ruins.

5. Full Extinction. Soaring levels of CO_2 produce levels of warming that make the entire Earth uninhabitable for humans and many other animal and plant species. Beyond the collapse of human sperm counts, other forces producing large-scale collapse and extinction include: widespread nuclear war; systems of artificial intelligence that escape human control; genetic engineering that produces an array of human species hostile to "ordinary" humans; loss of pollinating insects resulting in a mass extinction of plants and many animal species.[23] Efforts to prevent full extinction produce extreme genetic engineering to create designer humans with tolerance for high levels of heat and resistance to many diseases.[24] Weapons of bioterrorism could be created to hold humanity hostage, with threats to release pathogens unless there is a massive redistribution of wealth—and these pathogens could get out of control and complete the demise of humans on Earth.[25] Only fragments of life may remain, but from these, new forms of life could develop over tens of thousands or millions of years.[26]

In a world moving toward full collapse, two modes of adaptation seem likely to emerge:

1) *competitive* adaptation or a survivalist approach marked by groups in constant, violent struggle for the basics of life; and,

2) *compassionate* adaptation or a kindness approach, marked by eco-communities engaged in efforts for peaceful survival and the collaborative restoration of the local ecology.

Although a path of compassionate adaptation may be successful in the early stages of collapse, as the world becomes increasingly dominated by fierce struggles and conflicts over access to dwindling resources, it seems likely that communities of kindness would be attacked and overwhelmed by well-armed gangs who steal precious supplies of foods, seeds, plants, animals, and tools. Once extreme struggles for survival become widespread, it would be extremely difficult for people to come together with kindness and work cooperatively. A clear lesson emerges: *We should do everything we can to keep from descending into full collapse where wars for survival becomes normalized and initiatives for transformation are marginalized.*

To illustrate collapse leading to functional extinction, consider the example of Easter Island. With a mild climate and rich volcanic soil, Easter Island was a paradise covered by forests and filled with diverse animal and plant life when first settled by Polynesian colonists in approximately 500 A.D. As the islanders prospered, their numbers grew from a few hundred to an estimated 7,000 or more, and they rapidly consumed the resources of the island beyond its regenerative capacity. Archeological evidence shows that the destruction of the forests on Easter Island was well underway by the year 800—about 300 hundred years after people first arrived. By the 1500s, the forests and palm trees disappeared altogether, as people cleared land for agriculture, and used the remaining trees to build ocean-going canoes, to burn as firewood, and to build homes. Jared Diamond, professor of medicine at UCLA, describes how animal life was eradicated on Easter Island:

"The destruction of the island's animals was as extreme as that of the forest: without exception, every species of native land bird became extinct. Even shellfish were over exploited, until people had to settle for small sea snails. . . . Porpoise bones disappeared abruptly from the garbage heaps around 1500; no one could harpoon porpoises anymore, since the trees used for constructing the big seagoing canoes no longer existed. . . ."[27]

The biosphere was devastated beyond short-term recovery. With the forests gone, ocean fishing no longer possible, and animals hunted to extinction, people turned on one another. Centralized authority broke down, and the island descended into chaos with rival groups living in caves and competing with one another for survival. Eventually, according to Diamond, the islanders, "turned to the largest remaining meat source available: *humans*, whose bones became common in late Easter Island garbage heaps. Oral traditions of the islanders are rife with cannibalism." The only wild source of food that remained was rats. By 1700, the population had crashed to between one-quarter and one-tenth of its former level. When the island was visited by a Dutch explorer in 1722 (on Easter Sunday), he found a wasteland almost completely devoid of vegetation and animals. Cook described the islanders as "small, lean, timid, and miserable."[28]

The parallels between Easter Island and Earth are strong: Easter was an abundant island of life floating in a vast ocean of water. The Earth is an abundant island of life floating in a vast ocean of space. The meaning of Easter Island for us should be chillingly obvious, as Diamond concludes that Easter Island is Earth writ small:

> "When the Easter Islanders got into difficulties, there was nowhere to which they could flee, nor to which they could turn for help; nor shall we modern Earthlings have recourse elsewhere if our troubles increase. . . . If mere thousands of Easter Islanders with just stone tools and their own muscle power sufficed to destroy their environment and thereby destroyed their society, how can billions of people with metal tools and machine power now fail to do worse?"[29]

As Easter Island reveals, we humans have already demonstrated our ability, on a small scale, to devastate our biosphere irreparably and descend into functional collapse.

Future II: Authoritarianism

In this pathway, the extinction dangers of a whole-systems crisis for the Earth are recognized and, to reign them in, humanity trades personal freedoms and human rights for the safety promised by highly authoritarian communities or societies. Democracies are often cumbersome and slow, whereas authoritarian governments can act swiftly and with little concern for the views of the public. This streamlines decision-making and enables swift actions in the face of a crisis. The disadvantages of authoritarian governments include the oppression of minorities, the suppression of free association and expression, and the stifling of creative innovation. Authoritarian societies also have higher rates of mental illness, and lower levels of physical health and life expectancy.[30]

Digital dictatorships employ powerful computer technologies integrated across a range of areas (financial, social, medical, educational, employment, etc.) to tightly control their massive populations. In this pathway, the world avoids a devastating collapse by placing severe restrictions on nearly every aspect of life, thereby halting the descent into chaos. Trends of ecological, social, and economic breakdown are placed under strict control and stopped short of a ruinous collapse leading to functional extinction. A future of constraint and conformity lies ahead.

An often-cited example is China, which is creating a digital dictatorship using "social credit" scores combined with facial recognition systems and other technologies to monitor and control every person, with an array of punishments and rewards.[31] Cell phones and internet access are assigned unique numbers so they can be tracked. Transgressions that reduce one's public-trust score range from minor (jaywalking, playing video games too long) to major (promoting "fake news," "thinking infected by unhealthy thoughts," and criminal activity). Punishments range from public shaming (having your name and image posted publicly) to restricted work opportunities, diminished access to educational opportunities for yourself and/or your children, limited access to quality medicine, reduced internet speeds, and much more. Rewards include better job possibilities, better travel options (a plane instead of a bus), discounts on energy bills, easier access to hotels, and even better matches on computer dating sites. With artificial intelligence accelerating, punishments and rewards are continuously calculated for each individual in order to produce a highly monitored, regulated, and regimented society. Public opinion and discourse are tightly controlled by banning topics from news sources, promoting "pro-social themes," extensive monitoring of internet conversations, selective restricting of in-person gatherings of more than three people, and more. The result is a carefully watched, highly scrutinized, and controlled society that lives within ecological limits, but at the cost of a wide range of freedoms.

Importantly, China is not the only country moving forward with digital authoritarianism. The Chinese "Great Firewall" approach to the internet is spreading to a number of other countries, including Russia, India, Thailand, Vietnam, Iran, Ethiopia, and Zambia.[32] Even historically democratic nations such as the U.S. have a significant portion of the population—estimated to be roughly 20 percent of U.S. citizens in 2021—sympathetic to trading civil liberties for strongman solutions to secure law and order when confronted with social breakdowns.[33]

Although a number of nations have begun to consolidate authoritarian control over their populations, it is not clear if they can prevail over the long run in a world experiencing ruinous levels of climate change, water shortages, species extinctions, food shortages, and other ingredients of a world moving into a whole-systems collapse. Iron-fisted countries could break apart and give way to competing fiefdoms seeking to maintain authoritarian control at a smaller scale. Or worse: they could descend into full dictatorship, ruled by lone leaders with high narcissism and low compassion making decisions for all.

Future III: Transformation

A transformational pathway begins like the other two: Breakdowns continue and lead to a process of dynamic collapse. However, before collapsing into either functional extinction or the surrender of freedoms in authoritarianism, the people of the Earth could recognize the immense peril ahead and pull back from these two pathways and, instead, move forward on a path leading to a transforming world. Easier said than done! A transformational path requires far more than renewable energy, shifting diets, electric cars, and the equivalent of one-child families. We also require powerful forces for evolutionary uplift to transform a planetary systems crisis into a world that serves the wellbeing of all life.

Powerful, practical, and uplifting forces for building a transforming Earth are described at length in the last section of the book (Part IV), and are summarized below:

Seven Uplifting Forces

1. Choosing Aliveness—We shift from a mindset of separation and exploitation in a dead universe to one of community and care in a living universe. Living in the now with the direct experience of being alive becomes the source of meaning and purpose.

2. Choosing Consciousness—Paying attention to our movement through life with reflective consciousness or witnessing attention, we move outside the bubble of materialism and into compassionate participation with life.

3. Choosing Communication—Using tools of local-to-global communication, we grow a sense of local-to-global community and build a new consensus for our pathway into the future.

4. Choosing Maturity—Moving beyond a self-centered, adolescent mindset to mature regard for, and commitment to, the well-being of all life, we create the psychological foundation for a transforming future.

5. Choosing Reconciliation—Recognizing structural racism, extreme inequities of wealth and well-being, gender divisions, and "othering" in general, we search for healing and a higher common ground where cooperation and collaboration are awakened.

6. Choosing Community—Seeking security and a sense of belonging in a collapsing world, we begin rebuilding communities from the local level and re-discover feelings of being at home in the world.

7. **Choosing Simplicity**—Going beyond endless consumerism as the goal in life, we move toward grateful simplicity for being alive and choose to live with balanced regard for the well-being of all life.

There is no fantasy here. Each one of these uplifting forces is already widely recognized. The challenge is to energize and mobilize forces already present and available to us. The synergy of these two sets of changes—on the one hand, material changes (such as the proliferation of solar energy, new dietary patterns, a reduction in family size, new types of work, etc.) and, on the other hand, invisible changes (such as species maturation, consciousness, reconciliation, etc.) are vital for producing deep and lasting transformation. The intersection of these sets of changes will produce a dynamic and turbulent period of transition, as the evolutionary momentum of the past is gathered into a new dynamic for a transforming future. On the surface, this could appear to be a time of confusion and chaos; yet, deep currents of change would be at work, weaving and uplifting the world to a higher level of coherence, potential, and purpose.

Because a transformational path is assumed to emerge from a process of collapse, patience and persistence will be vital for evolutionary uplift to blossom visibly in the world. Although this pathway is profoundly demanding—for example, calling forth a new level of maturity, reconciliation, and consciousness from humanity—it is already within our current capacity to choose.

It is helpful to recognize the many areas where humans have long been working together successfully.

- *Weather*—The world weather system merges information from more than one hundred countries every day to provide weather information globally.
- *Health*—Nations around the globe have cooperated to eradicate diseases such as smallpox, polio, and diphtheria.
- *Travel*—International aviation agreements assure the smooth functioning of global air transport, while global cooperation has enabled the International Space Station to be built by a consortium of nations.

- *Communications*—The International Telecommunications Union (ITU) allocates the electromagnetic spectrum so that television signals, cellular phones, and radio signals are not overwhelmed with noise.
- *Justice*—Global ethics are emerging as world courts and tribunals hold heads of state accountable for policies of genocide, torture, and crimes against humanity.
- *Environment*—Despite lagging climate action, nations of the world have reached important agreements on ecological concerns, such as banning CFCs that damage the atmosphere's ozone layer.

These examples of successful collaboration among the human community provide an important context for looking ahead—they illustrate humanity's capacity to rise to a higher maturity and to work together effectively.

It is helpful to look at the three primary pathways side-by-side to see their similarities and differences. What most differentiates these three possible futures is not the underlying trends, but the overlying choices we humans make. Because there is no single most likely future, the pathway that prevails will depend on which we consciously choose—or unconsciously surrender. Therefore, a pathway of uplifting transformation is not a prediction; instead, it is a plausible description of collective choice and shift in consciousness we could realize as a global society in response to breakdown and dynamic collapse.

One of our most important capacities as a species is the ability to look ahead, anticipate what might unfold, and then respond swiftly. If we can use our collective imagination to envision how we are creating an uninhabitable Earth, then we don't have to manifest that future in physical reality in order to learn its lessons. We can internalize the teachings and insights from an imagined future and consciously choose a different path ahead. We have already begun to vividly imagine futures we don't want to inhabit. In turn, we don't need to wait for global warming to melt the icecaps and flood the world's sea-coast cities before we wake up and decide this is not a future we want. We don't need to kill off a million different species of animals and plants before we decide that

Figure 2: Three Pathways for Humanity

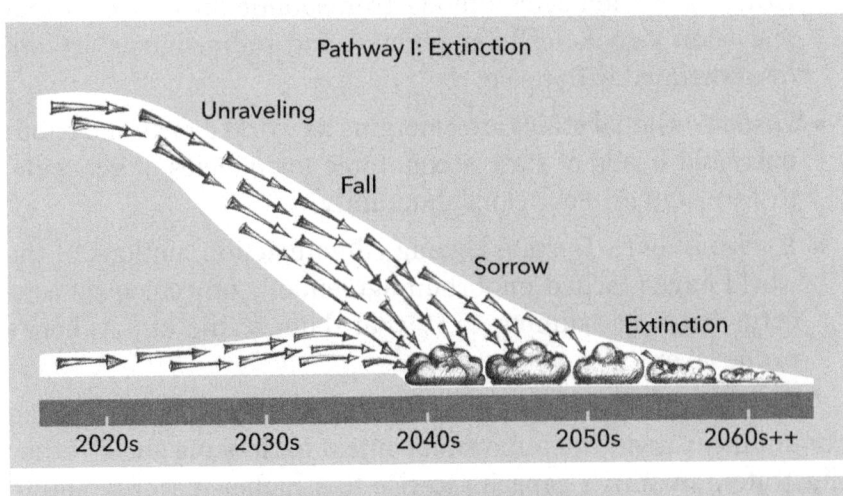
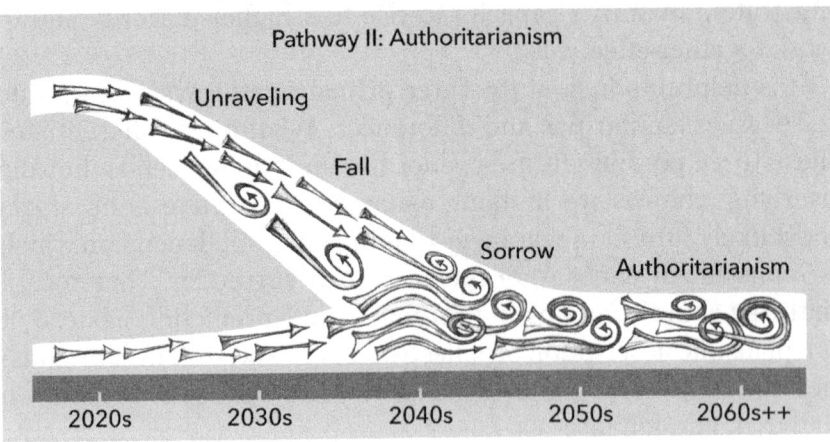
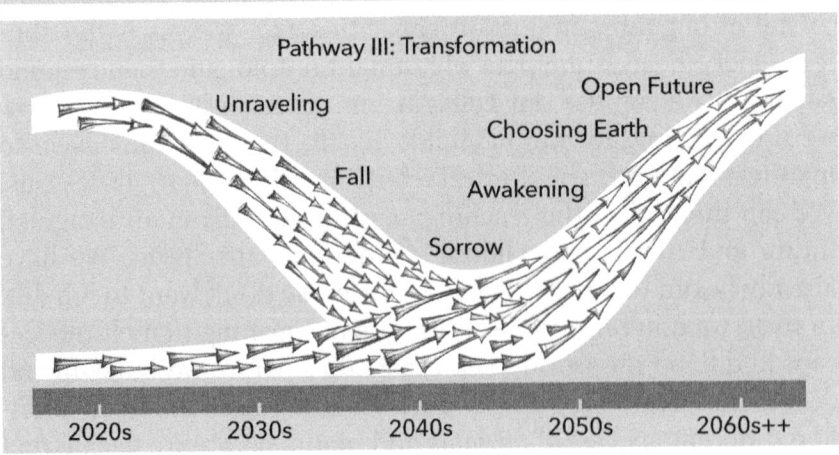

an impoverished and barren biosphere is not a future we choose. We don't need to surrender to authoritarian rule and a digital dictatorship before we decide that human freedoms for conscious evolution are precious beyond measure. If we mobilize our collective imagination and visualize more clearly the pathways ahead, we can consciously orient toward a different future—*now,* rather than after years of delay and distraction.

PART III

Stages of Initiation and Transformation

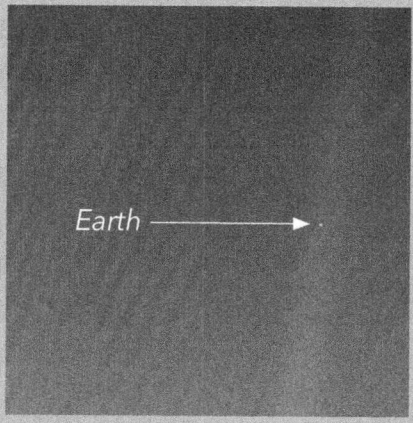

The Earth seen by the Voyager spacecraft from nearly 400 billion miles away

"Look again at that dot. That's here. That's home. That's us. On it everyone you love, everyone you know, everyone you ever heard of, every human being who ever was, lived out their lives. The aggregate of our joy and suffering, thousands of confident religions, ideologies, and economic doctrines, every hunter and forager, every hero and coward, every creator and destroyer of civilization, every king and peasant, every young couple in love, every mother and father, hopeful child, inventor and explorer, every teacher of morals, every corrupt politician, every "superstar," every "supreme leader," every saint and sinner in the history of our species lived there—on a mote of dust suspended in a sunbeam."

—Carl Sagan

Scenario of Transformation

We now move from the three pathways to consider a "transformational" future in depth. The other two paths of "extinction" and "authoritarianism" are relatively clear as they are already visibly emerging in the world. However, a transformational future is different because it represents an evolutionary advance into the unknown. Because we have never ventured there before, we don't have predetermined ideas what a transformational pathway looks like. It builds on the combined power of uplifting forces we recognize individually, but have not imagined converging into a collective force for evolution. To offer a glimpse into a transformational view of the future, here is a paragraph taken from my 2009 book *The Living Universe*:

> "The suffering, distress, and anguish of these times will become a purifying fire that burns through ancient prejudices and hostilities to cleanse the soul of our species. I expect no single, golden moment of reconciliation to descend upon the planet; instead, waves of ecological calamity will reinforce periods of economic crisis, and both will be amplified by massive waves of civil unrest. Instead of a single crescendo of crisis and conflict, there will likely be momentary reconciliation followed by disintegration, and then new reconciliation. In giving birth to a sustainable world civilization, humanity will probably move back and forth through cycles of contraction and relaxation. Only when we utterly exhaust ourselves will we burn through the barriers that separate us from our wholeness as a human family. Eventually, we will see that we have an unyielding choice between a badly injured (or even stillborn) planetary civilization and the birth of a bruised, but relatively healthy, human family and biosphere. In seeing and accepting responsibility for this inescapable choice, we will work to discover a common sense of reality, identity, and social purpose. Finding this new common sense will be an extremely demanding task. Only after we have exhausted all hope of partial solutions will we be willing to move forward with an open mind and heart toward a future of mutually supportive development. Ultimately, in moving through our initiation, we can grow from our adolescent ways as a species into our early adulthood and

consciously take responsibility for our relationship with the Earth, the rest of life, and the universe."[34]

This paragraph does not describe with any detail the nature of transformational changes that lie ahead. To develop a more robust scenario of the future, below I describe each decade in three different ways:

1. A **summary** of the decade. It is easy to get lost in the detailed information so the summary gives an overview of the decade.

2. A review of the major **driving trends** in each decade. This is the hard, factual information from the most trusted sources I could find to develop a detailed understanding of the major challenges that lie ahead. Driving trends provide the "skeleton" or analytical framework for the scenario.

3. A scenario or **story** describing how the decade unfolds. This is the "flesh" of a more subjective description of how the decade develops. Detailed trends are woven together into a realistic narrative about the future.

Drawing on the best scientific estimates available, I have identified eight driving trends common to each decade:

1. Global warming and climate disruption
2. Water scarcity
3. Food scarcity
4. Climate refugees
5. Species extinction
6. World population
7. Economic growth/breakdown
8. Economic inequities

Where futures research often considers only a few driving trends, I consider all eight and how they are likely to interact with one another over the coming decades. Then I develop an additional seven *uplift factors*—the "flesh" that fills out the skeletal descriptions. Bringing together these fifteen driving factors, a scenario emerges that is rich with detail. This approach does not guarantee "right answers" about the future; however, it does offer a disciplined approach for developing a realistic view of an uplifting pathway that can emerge from these dark decades.

It is important to acknowledge that dividing this scenario into ten-year increments is rather arbitrary. The world is a messy and complex place that does not divide its progression into neat and convenient decades of development. Additionally, we have entered a turbulent and chaotic time of planetary transition that will contain wildcards—such as the sudden emergence of the global Covid pandemic—that can put otherwise plausible expectations into a tailspin. So, there is good reason to be cautious about dividing the future into discrete decades.

Because scientific confidence in trend data diminishes as we look further into the future, the early decades are more heavily weighted with scientific data and analysis. As mentioned previously, *all three pathways—extinction, authoritarianism, and transformation—begin with these same driving forces.* The difference between them is not in the early trends, but in the choices the human community makes in response to those trends. *A transformational future unfolds only because we raise our heads and awaken our hearts to follow a higher purpose and potential as a species.*

Exploring a transformational scenario is a demanding exercise in social imagination that requires compassion, persistence, and patience. *This is difficult work.* We must mobilize every faculty we have to develop a clear picture of the future—one that includes sorrows and losses as well as powerful uplifting factors such as collective maturation and awakening that can transform unyielding adversity into realistic opportunity. Although exploring the next fifty years is very challenging, it offers the potential for visualizing a profound initiation and rite of passage for our species.

I want to pause for a moment and acknowledge your courage in choosing to read this book. You read on behalf of all life. I assume you are a person with a curious intelligence and a compassionate heart. I assume you care for life, for people, for nature, and for the Earth. I assume you feel intuitively how life in the future calls to all who are awake in the present to bear witness to what is now unfolding on Earth. To step up as a witness to our time of unprecedented transformation is a gift to the future. Until recently, few people were aware that a dynamic collapse of human civilization is underway, creating a profound initiation for our species. Today, we can consciously recognize an initiation is in progress—and this knowledge can make a tremendous difference in choosing our path forward. I honor both your feelings of loss as well as your gratitude for the life that continues. I respect your willingness to see what is unfolding. By doing so, you are contributing to a new kind of human that can serve the wellbeing of all life. Thank you for being a faithful servant to our transforming future.

2020s: The Great Unraveling— Breakdown

Summary

In the 2020s, the great transition gets underway as humanity awakens to the unyielding fact that we confront a profound world crisis. We gradually recognize that instead of a single problem to be fixed, we confront a whole-systems crisis that requires deep changes in how we live on Earth. Collectively, we do not come to this understanding quickly or easily. Humanity enters this pivotal decade with deep divisions. Slowly, a minority of people awaken to the reality that we face a whole-systems crisis that goes far beyond climate disruption.

In this decade, global warming is increasing droughts, fires, floods, and intense storms around the world. Actions to cope with rising CO_2 are underway, but the pace of innovation lags far behind what is required to stabilize global temperatures. We are on the road to climate catastrophe. Water scarcity is a source of stress for nearly half of the world's population. Aquifers are being pumped dry in the U.S., India, and elsewhere around the world. Several million more people each year become climate refugees as they seek to move into more resource-favored areas. Animal and plant species are stressed, unable to swiftly migrate in response to the rapid rate of climate change. Economic supply chains are breaking down.

Institutions of all kinds (economic, political, academic, healthcare, etc.) are slow to make changes. Most leaders focus on protecting their wealth, power, status, and privilege. Leaders are more concerned with perpetuating their institutions than in protecting the well-being of all life. A profound loss of trust in leadership continues to grow among the world's younger generations. A majority of younger people feel "doomed" and that their long-term future has been abandoned in favor of short-term gain by the older generation.

Challenges to the mindset of materialism, consumerism, and capitalism are growing, but are largely ineffectual given the economic and political power of the wealthiest individuals. At the global level, disparities in wealth are extreme: the top (richest) 10 percent of the world's population has 76 precent of the wealth and the bottom 50 percent of the population has only 2 percent of the wealth. In other words, 10 percent of the world is taking three quarters of the total wealth and leaving the bottom half of the world's population with only a tiny percent of the wealth.[35] Importantly for climate change, these inequalities reflect more than disparities in economic well-being; they also reflect great differences in CO_2 emissions. The wealthy are responsible for emitting a disproportionate amount of carbon. It looks increasingly doubtful whether our world can work together as an integrated and cooperative whole with such extreme differences. A global wealth tax and carbon tax is important if we are to make a transition to a low-carbon world and provide adequate health care, education, and restore ecological health for the planet. Although the need for greater fairness is enormous, resistance is even stronger. It looks likely that the economic system that supports these profound inequities will collapse under the weight of this dysfunction. It's just not sustainable.

The communications revolution continues at a rapid pace with high-speed networks coming into wide use in the U.S. and growing globally. Two-thirds of the world's citizens have access to the internet at the beginning of the decade, growing rapidly to three-quarters by the end of the decade. However, the consumer-oriented content of communication generally promotes a more adolescent, self-centered mindset focused on the short-run.

Overall, in this decade, conflicts grow as people increasingly withdraw into groups identified by race, ethnicity, religion, wealth, and political orientation. Despite growing breakdowns, the primary concern is returning to the old normal and continuing business-as-usual.

Review of Major Driving Trends in the 2020s

- **Global Warming**: A 1.2° Celsius rise of global warming (roughly 2° Fahrenheit) by 2020 provides clear evidence that major climate disruption is underway. Scientists are concerned that a 1.5°C increase will produce much more climate instability than previously thought.[36] Alarming scientific projections estimate that a catastrophic temperature rise in the range of 3°C will develop by the end of the century.

  The implications for global warming are horrendous: For example, a 2019 IPCC *Special Report* recognized that half of the world's megacities, with almost two-billion people, are located on vulnerable coasts. Even if global temperature rise is restricted to just 2°C, scientists expect the impact of sea-level rise to cause several trillion dollars of damage annually and result in many millions of people migrating from coastal areas.[37] The special report offered this grim picture of the long-term future:

  "We have simply waited too long to reduce emissions and will be forced to grapple with impacts that can no longer be avoided. However, the difference between sharply reducing emissions and continuing along the "business-as-usual" pathway is stark: Under a low-emissions scenario, managing the impacts of climate change will be expensive, but possible; doing nothing will result in unmanageably catastrophic effects."[38]

  Sea-level rise will continue for hundreds, perhaps thousands, of years, even if emissions are reduced to zero now.[39] Despite clear warnings of catastrophe, CO_2 emissions continue to grow.[40] This raises fears that we may create a "hothouse Earth" condition unlike anything in human experience.[41]

  In addition to temperature rise producing warming oceans, shrinking ice sheets, and ocean acidification, global warming also brings new weather extremes—storms, rain, floods, and droughts—that severely impact agriculture and habitats.⁴² All of these changes are expected to intensify throughout the 21st century and beyond.

Global warming has a direct impact on human health, as well. A report from the World Health Organization states: "The climate crisis is a health crisis... that exacerbates malnutrition and fuels the spread of infectious diseases such as malaria. The same emissions that cause global warming are responsible for more than one-quarter of deaths from heart attack, stroke, lung cancer, and chronic respiratory disease."⁴³

- **Pandemics**: For a number of reasons, pandemics—diseases that spread worldwide—are more likely to emerge from conditions produced by global warming.

  1. As frozen regions of the Earth begin to thaw due to global warming, they release viruses that have been locked away for tens of thousands of years. During the preceding ice ages, both humans and other animals may have lowered their disease resistance and become much more vulnerable to their infections.

  2. New pandemics emerge as economic advances support dramatic population growth and lead to large human populations living in close proximity with wild animals, enabling diseases to jump more easily to humans.

  3. With technological advances and high mobility, a more rapid mixing of people and wild animals around the Earth gives viruses a way of making their journey around the world swiftly. The scope and speed of modern human travel makes quarantines nearly impossible to implement and enforce.

  4. Technological advances create the possibility of terrorists manufacturing or bio-engineering pathogens as bio-weapons to produce pandemic threats.

Pandemics—such as the coronavirus—are likely to become a recurring disturbance in a rapidly warming world.⁴⁴ While pandemics are unlikely to be the catalyst for a global civilizational collapse, they do reveal the vulnerability of our tightly interconnected social and economic systems. They also offer a compelling example of the need for mature, planetary collaboration. Covid has awakened humanity to our collective vulnerability, and demonstrates how a vigorous response by only a few nations will not be adequate. In our highly mobile world, new variants of the virus can spread around the planet within weeks. Stopping the virus before new variants can emerge and spread would require nearly all humans to be vaccinated at roughly the same time—a global response to a global threat. Covid is awakening an Earth-scale collective consciousness as we grapple with how to respond. However, key differences exist between the climate crisis and pandemics. Although pandemics reveal that we are all connected in the Earth's web of life, they are generally perceived as a relatively discrete, nearby, immediate, and personal threat to one's self and family. In comparison, climate disruption is a more complex, deeply interconnected, distant, vague, and general in its threat to society and the larger economy. Actions required to respond to the climate crisis are not simple and the benefits of those actions are less certain and less immediate. Ambiguity and uncertainty make a unified response and decisive climate action much more difficult. Despite these differences, the coronavirus pandemic is making an important contribution to humanity's awakening to the reality of living in a tightly interdependent world.

- **Water Scarcity**: Although huge oceans cover the Earth, only three percent of the planet's water is fresh, and much of that is inaccessible—with over two-thirds of fresh water locked up in icecaps and glaciers, and nearly all the rest found in ground water. Only three-tenths of one percent of all fresh water in the world is found in surface lakes and rivers. Given the enormous increase in world population with water-intensive ways of living, water is already becoming a scarce resource. In 2020, an estimated 30 percent to 40 percent of the world experienced water scarcity, and by 2025 an estimated three-billion people will live in areas plagued by water scarcity, with two-thirds of the world's population living in water-stressed regions.[45] In 2019, "844 million people, 1 in 9, lacked access to safe water and 2.3 billion people, 1 in 3, lacked access to a toilet."[46] More than two-billion people live in countries experiencing high water stress, and about four-billion people experience severe water scarcity during at least one month of the year. Stress levels will continue to increase as demand for water grows and the effects of global warming intensify.[47]

- **Food Scarcity**: "In 2019, a little over 800 million people suffered from hunger, corresponding to about one in every nine people in the world."[48] Despite significant improvements in previous decades, the food prospect for the future is grim due to climate disruptions.[49] To illustrate the predicament, "According to UNICEF, 22,000 children die each day due to poverty. And they die quietly in some of the poorest villages on Earth, far removed from the scrutiny and the conscience of the world." Around 27 percent of all children in developing countries are estimated to be underweight or stunted."[50] Global demand for food will more than double over the coming half-century as we add roughly another two to three-billion people. A central issue in the coming half century is whether humanity can achieve and sustain such an enormous increase in food production.[51] Another study found that:

  "Decisions made in the next few decades will have huge ramifications for the future of our planet, and getting our food systems right is at the heart of this. Current practices are contributing to the problem, all in an effort to produce the record amounts of food needed to feed our global population . . . it was this very progress that contributed to large-scale land and water degradation, biodiversity losses, and increased greenhouse gas emissions. Now, the productivity of 23 per cent of global land has declined, and about 75 percent of freshwater is used just for agriculture."[52]

- **Climate Refugees**: Between 2008 and 2015, an average of 26.4 million people per year were displaced by climate or weather-related disasters, according to the United Nations.[53] Tens of millions of people were on the move in 2020.

- **Species Extinction**: By the end of this century, a UN report concludes, more than one million species of plants and animals are at risk of extinction—many of which are predicted to be pushed into extinction within just a few decades. Robert Watson, a British chemist who served as the panel's chairman, stated, "The decline in biodiversity is eroding the foundations of our economies, livelihoods, food security, health, and quality of life worldwide."[54] The integrity of the biosphere is being devastated and losses include insects, birds, mammals, and reptiles, as well as fish. The overall outlook is very grim.

The world's **insects** are hurtling down a path to extinction, threatening a "catastrophic collapse of nature's ecosystems" according to the first global scientific review.[55] The analysis found that more than 40 percent of insect species are declining and a third are endangered. The rate of extinction for insects is eight times faster than that of mammals, birds, and reptiles, and is so great that, "Unless we change our ways of producing food, insects as a whole will go down the path of extinction in a few decades. . . . The repercussions for the planet's ecosystems are catastrophic, to say the least."

Bees are also disappearing at an alarming rate due to the excessive use of pesticides in crops and the spread of certain parasites that reproduce only in bee colonies. *The extinction of bees could mean the end of humanity. If bees didn't exist, it is hard to imagine humans surviving.* Out of the 100 crop species that provide 90 percent of our food, 35 percent are pollinated by bees, birds, and bats.[56]

Another study found that **birds** are vanishing from North America: The number of birds in the United States and Canada has declined by three billion, or 29 percent, over the past half-century.[57] David Yarnold, president of the National Audubon Society, called the findings "a full-blown crisis." Kevin Gaston, a conservation biologist, said that new findings signal something larger at work: "This is the loss of nature." "The skies are emptying out. There are 2.9 billion fewer birds taking wing now than there were 50 years ago."[58] The analysis, published in the journal *Science,* is the most exhaustive and ambitious attempt yet to learn what is happening to avian populations. The results have shocked researchers and conservation organizations.

The **ocean's** eco-system is being devastated, with marine life declining by 49 percent between 1970 and 2012. Overfishing and pollution are producing an "unprecedented" marine extinction. A major report found that every species of wild-caught seafood—from tuna to sardines—will collapse by the year 2050. "Collapse" was defined as a 90 percent depletion of the species baseline abundance.[59] Another report warns that hunting and killing of the ocean's largest species will disrupt ecosystems for millions of years.[60]

Here is how the Center for Biological Diversity describes the overall extinction crisis:

"Wildlife populations are crashing around the world.... Our planet now faces a global extinction crisis never witnessed by humankind. Scientists predict that more than one million species are on track for extinction in the coming decades. Wildlife populations around the world are crashing at alarming rates and with distressing frequency.... When a species goes extinct, the world around us unravels a bit. The consequences are profound, not just in those places and for those species, but for all of us. These are tangible consequential losses, such as crop pollination and water purification, but also spiritual and cultural ones. Although often obscured by the noise and rush of modern life, people retain deep emotional connections to the wild world. Wildlife and plants have inspired our histories, mythologies, languages, and how we view the world. The presence of wildlife brings joy and enriches us all—and each extinction makes our home a lonelier and colder place for us and future generations. The current extinction crisis is entirely of our own making."[61]

- **World Population**: At the beginning of the 2020s, world population is roughly 7.8 billion.[62] Although population projections to the end of the century are difficult, a median estimate of total world population in 2100 is approximately 11 billion. Rough estimates are that, by 2100, the top five most populous countries will be: India, with 1.2 billion people, China with 1 billion, Nigeria with nearly 800 million (comparable to the entire population of Europe in 2010), the U.S. with 450 million, and Pakistan with 350 million.[63]

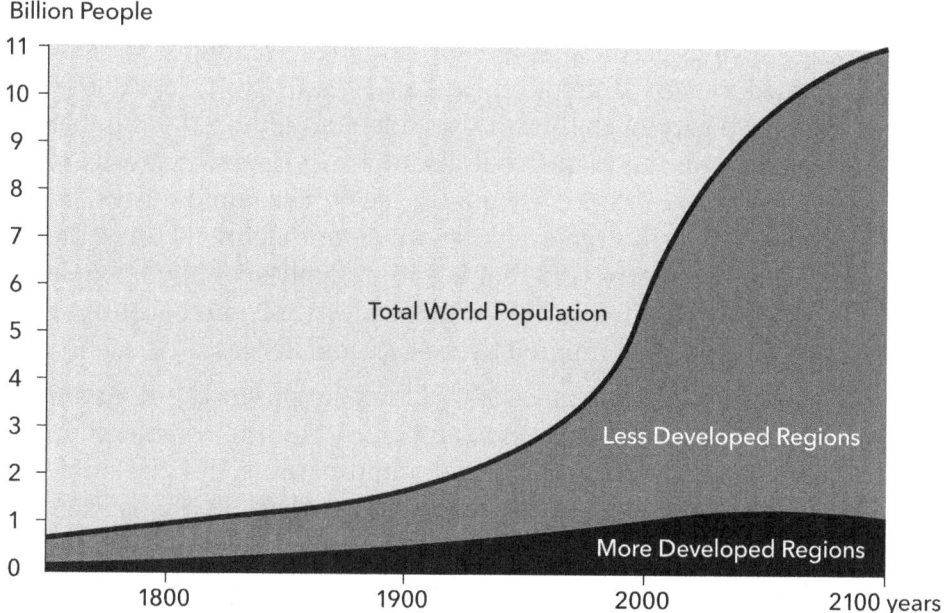

Figure 3: World Population Growth: 1750-2100[64]

Less developed regions: Africa, Asia (excluding Japan), Latin American and the Caribbean, and Oceana (excluding Australian and New Zealand).

More developed regions: Europe, North America (Canada and the United States), Japan, Australia, and New Zealand.

An estimate of global population of roughly 11 billion is far from certain—particularly if deep and rapid shifts to sustainable ways of living are not adopted. Given current food production capacity and water resources, the Earth can support roughly nine billion people *if resources are shared equally*. However, with agricultural productivity falling due to global warming and water scarcity, the carrying capacity of the Earth is declining. In addition, much depends on the consumption patterns of developed nations relative to the rest of the world. If the entire world consumed at the same level as the United States, the Earth could support roughly 1.5 billion people. With middle-class European lifestyles, the carrying capacity grows to roughly two billion people.[65] The Earth supports U.S. levels of consumption only because people in the U.S. are drawing down the "savings account" of non-renewable resources, such as fertile topsoil, clean drinkable water, virgin forests, undiminished fisheries, and untapped petroleum.

Our "savings account" is already running low, and we are now obliged to live within our means as a species. In turn, the carrying capacity of the Earth depends not only on the number of people on the planet, but also their levels and patterns of consumption. In the early 2020s, the human community consumes the Earth's renewable resources at roughly 1.6 times the sustainable rate[66] — and that is with roughly six billion people involuntarily living in "low-carbon lifestyles" and consuming next to nothing compared to the U.S. middle class.

Given the great reluctance of wealthier nations to sacrifice their high-consumption lifestyles, and given that the consumption footprint of the Earth is rapidly approaching nearly double what the Earth can provide on a long-term basis, it seems likely that a dramatic die-off in human numbers will occur. Is the great suffering that will result *unavoidable?* Will it take such a catastrophe to push people in developed nations to make needed changes in their levels and patterns of consumption? How much pain and suffering are required for humanity to turn to a new equilibrium and fairness in global consumption?

- **Economic Growth/Breakdown**: Secure networks of economic activity around the world are beginning to break down. The global economy is unraveling, supply chains are coming apart, and the flow and delivery of goods are increasingly unpredictable. Key materials (ranging from wood products to computer chips) are becoming scarce, ports are becoming congested, shipping costs are growing, and deliveries to customers are becoming unreliable.

  Experts widely agree that roughly 70 percent of economic activity in the U.S. is connected with the production of consumer goods, which is understandable for a consumer-based economy.[67] Numerous studies conclude, "Emissions are a symptom of consumption and unless we reduce consumption, we'll not reduce emissions."[68] Therefore, future economic growth likely will be diminished by the urgent need to reduce carbon emissions and, therefore, the need to reduce overall consumption levels. "It doesn't matter if you're in a hot or a cold climate, a rich country or a poorer one—an unchecked Earth systems crisis is going to devastate the economy. This research comes as the United Nations says that climate impacts are happening faster, and hitting harder, than anticipated."[69] The risks associated with climate change are not being integrated into pricing and this, in turn, reduces incentives needed to decrease emissions—an economic error with catastrophic consequences.[70]

"The next two decades will be decisive. They will determine whether we suffer severe and irreversible damage to livelihoods and the natural world or whether, instead, we set off on a more attractive path of sustainable and inclusive economic development and growth.... If we go on emitting greenhouse gases at current rates for the next two decades, then it is likely that we will far exceed a 3°C increase.... A rise of 3°C would be extremely dangerous, taking us to a temperature we have not seen on this planet for around three million years.... A warming of this magnitude could transform where we could live, severely damage livelihoods, displace billions of people and lead to severe and extended conflict."[71]

- **Economic Inequities**: It doesn't matter how we look at it: global inequality of wealth and income is getting worse, much worse. In 2017, the world's six richest men were as wealthy as half of humanity![72] Six individuals have as much wealth as 3,600,000,000 of the world's poorest people. Equally stunning is the estimate that the richest one percent of the world's population has more wealth than the rest of the world's population combined.[73]

The astonishing inequity in the United States is revealed by the astounding fact that tax rates for the wealthiest are lower than for any other income group: "For the first time on record, the 400 wealthiest Americans in 2019 paid a lower total tax rate—spanning federal, state, and local taxes—than any other income group."[74] As long as a wealthy elite has the power to set the rules to their own advantage, inequality will continue to worsen.[75]

A powerful way to visually represent the unfairness and injustice of global income distribution is by looking at the shape of the following figure, where income for the world is divided into five groups, each representing 20 percent of the world from low- to high-income.[76] The long, thin portion of the outline (akin to the stem of a champagne glass) represents the annual income of a majority—approximately 60 percent of the people in the world. The portion where the stem begins to widen out represents the income of the next 20 percent—the global middle-class. The widest portion illustrates the income received by the world's wealthiest 20 percent. Just by looking, it is apparent that the human family is made up of a huge, impoverished class, a small but growing middle-class, and a very small and extremely wealthy elite.

These inequities have major consequences for the Earth's climate. Almost 50 percent of global carbon emissions are generated by the activities of the wealthiest ten percent of the global population. In stark contrast, the poorest 50 percent of the global population are responsible for around only ten percent of global carbon emissions yet live overwhelmingly in the countries most vulnerable to climate change.[77] Given these immense disparities, climate adaptation is already a profound issue of social justice.

Climate "justice" means that those who are least responsible for climate change should not be the ones to suffer its gravest consequences.[78] Yet structural inequities, often based on race, mean that communities of color will continue to be hit "first and worst" in the climate crisis.[79] To right this imbalance, a high priority should be the imposition of a per-capita carbon emissions limit on the top ten percent of global emitters (roughly equivalent to that of an average European citizen). If this were done, global emissions could be reduced by one-third in a matter of a year or two!

Historically, great disparities of wealth have been a consistent precursor to dramatic social breakups and violent change. If humanity wants to avoid deep civil conflict, then it is vital to recognize how the current economy is not working for the

Figure 4: Global Wealth Distribution

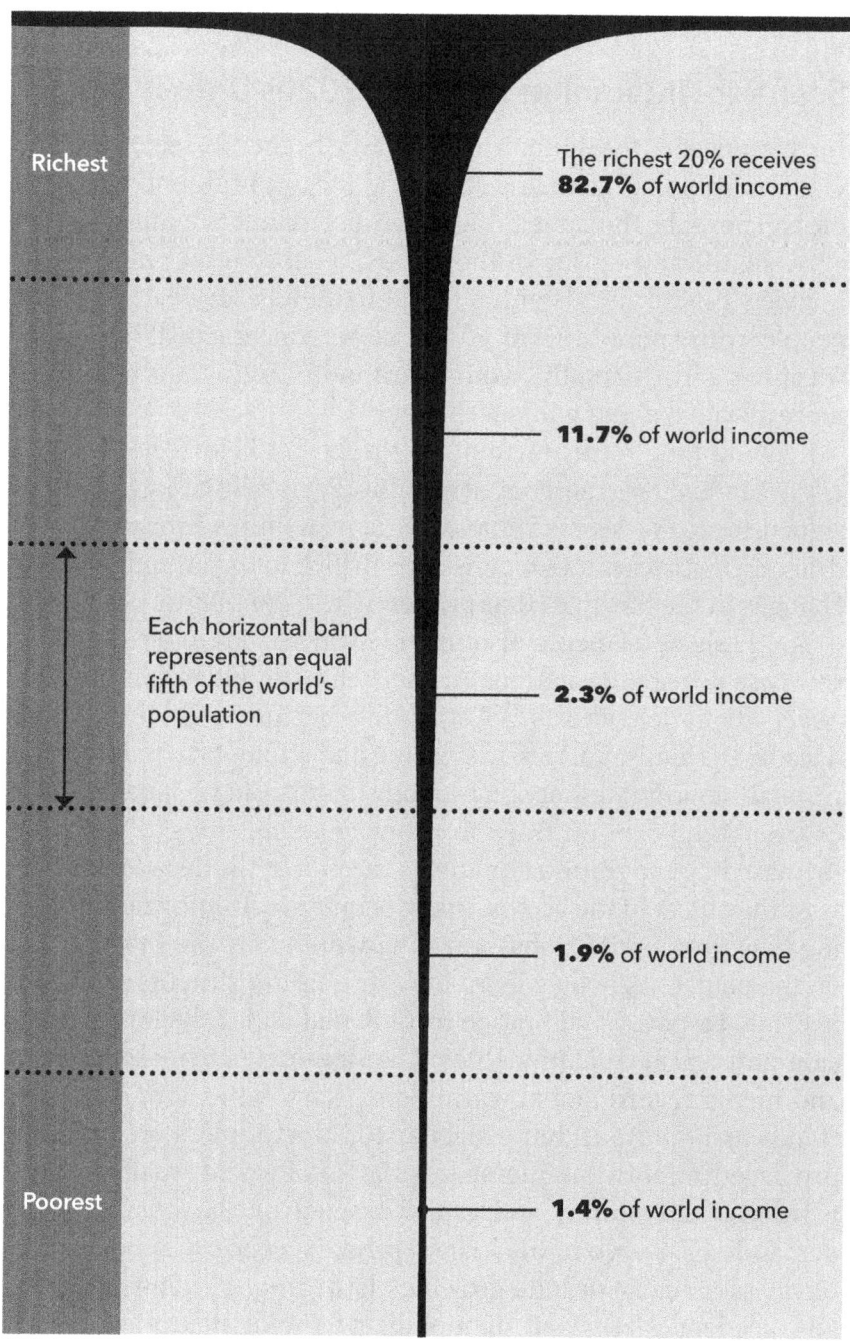

benefit of the majority. A voluntary shift in favor of a much more equitable distribution of wealth is a very wise course of action.

Scenario: Imagining How the 2020s Unfold

In this decade, the human community begins to recognize that global warming is changing the world in such profound ways that life will never be the same. Although concerns about climate change grew significantly prior to the 2020s, a substantial minority did not view this as an existential threat to human survival.[80] Overall, people with more education are more concerned about global warming and, generally, women are more likely than men to be more alarmed about climate change.[81]

Keeping long-term warming of the planet below the target of 1.5°C (or 2.7°F) — the goal set in the Paris Climate agreements signed in 2015 — seems impossible as it requires immediate and dramatic cutbacks in CO_2 emissions, which in turn require radical changes in the lifestyles that produce these emissions.

The Paris agreements also include ways for developed nations to assist developing nations in their climate mitigation and creative adaptation efforts.[82] Yet at the beginning of this pivotal decade, CO_2 emissions are increasing and attempts to reduce them through coordinated actions among nations have failed. Global CO_2 emissions are on track to produce a dangerous, 2°C (3.6°F) increase in temperature as early as the end of the decade.

At the outset of the 2020s, many people are ill-informed regarding how profoundly global warming will impact the future of life on the planet. As people learn how serious our situation will soon become, responses will range from denial and disbelief to confusion and alarm. Wealthy elites who dominate business, politics, and media regard global warming, species extinction, and other trends as important, but exaggerated. Most leaders are among a privileged minority immersed in the comforts of wealth, status, privilege, and power, and are distracted by the busyness and demands of everyday life. Their primary concern is continuing business-as-usual despite growing alarm among scientists, youth, and academics. Instead of mobilizing for dramatic action and innovation, privileged elites seek only gradual adjustments that don't disrupt the status quo.

Mainstream media strongly support the social trance of consumerism with endless entertainment—sports, reality television, movies, video games, and celebrity gossip—that glamorizes consumer lifestyles and deflects and numbs social attention.

Although climate disruptions and a cascade of other difficulties are growing, influential leaders soften claims of an intertwined, whole-systems crisis. Instead, problems such as climate change are portrayed as:

- Not as important as other issues, such as jobs and healthcare.
- Not as urgent or immediate as claimed, so we have ample time to respond.
- Not as large in scope as claimed.
- Not as difficult to remedy as claimed; assuming technology will fix many of the problems.
- Not a whole-systems crisis; rather, these are individual problems that can be solved one at a time.
- Not a graspable problem individuals can solve: "What can I do? I'm only one person."
- Not my responsibility: "I did not create this mess, so why ask me to clean it up?"

The "soft denial" of many leaders combines with a pervasive sense of helplessness. Understandably, business-as-usual persists and mainstream institutions respond with half-hearted measures that do little to slow a relentless advance toward a disastrous future. Nonetheless, a small fraction of people is adapting how they work and live.

The United States—the world's leading consumer nation—illustrates the difficulty of dealing with transition in a constructive way. Rev. Victor Kazanjian of the United Religions Initiative describes how the U.S. is a society of grievance unable to accept our fate and grieve over the changes required of us. He writes:

> ". . . much of what underlies anger, rage, and violence is grief—a sense of loss upon loss upon loss. But in our culture, we don't have much space for grief. Grief when unaddressed

becomes grievance. We are in a culture of grievance. Grief being expressed as blaming the other. We have to address deep grief."

Despite great resistance, by the middle of the 2020s, the disruptions of climate and natural systems are becoming so great they begin to break the consensus trance of consumerism, distraction, and denial. Climate emergencies multiply and awaken a growing recognition that Earth-scale challenges are underway. Complacency gives way to growing alarm as seasons around the planet are so disrupted that food production is compromised, producing regions of severe famine and civic unrest.

The overarching challenge of the 2020s is to awaken our social imagination to the imperative of making extraordinary changes in how we live on the Earth and to acknowledge that an entirely new approach to the future is required if CO_2 emissions are to be reduced and brought under control.

- Gradually, the more materially privileged begin shifting from overconsumption toward lifestyles of "voluntary simplicity," while the impoverished continue with involuntary simplicity and a daily struggle for survival.
- There is a growing outcry against extreme inequities of wealth and well-being. A consensus grows to "tax the billionaires" to fund healthcare safety nets, social security systems, and infrastructure repair.
- For the more affluent, diets begin to shift toward vegetarianism, transportation shifts toward electric vehicles, homes become more energy efficient, and work shifts toward decreased environmental impact and increased social contribution and meaning.
- Ecological lifestyles move from a fringe movement for a few to a wave of experimentation for mainstream culture. Low-carbon, materially simple, and experientially rich lifestyles become more widespread. For most, it's a relatively superficial way of "going green."
- Materialism and consumerism are increasingly called into question, as people challenge cultures of aggressive advertising, they declare: we are more than consumers to be entertained;

we are citizens of the Earth who want to participate in creating a more sustainable future.
- New configurations of economic activity begin to emerge that emphasize local resilience, skill sets, and patterns of work.

By the end of the decade, a transition in culture and consciousness gets underway, primarily in rich countries where people have the luxury of looking beyond daily survival. A clear-headed understanding grows that new approaches to living are essential—but actions are seldom commensurate with need.

For the past few decades, a mindfulness revolution has been growing throughout the planet. Relatively small, but significant, numbers of people are developing the skills of reflective consciousness—the capacity to simply witness their lives and to live with less reactivity and greater maturity. A small-but-significant fraction of humanity is beginning to wake up and grow up. With reflective consciousness, we more clearly witness ecological crises, poverty, overconsumption, racial injustice, and other conditions that have divided us in the past. With a more reflective perspective, we begin to develop collective understanding that serves the well-being of all. Reflective consciousness provides the invisible glue to begin bonding the human family into a mutually appreciative whole, while simultaneously honoring our differences.

With a witnessing consciousness growing, people recognize that the whole-systems crisis is a *communications* crisis, and this gives rise to diverse communication initiatives—ranging from living-room conversations to dialogues and conferences among leaders in business, government, media, education, religion, and more. These are important, but painfully inadequate. The scale of communication does not match the scale of the challenges we face. People recognize that the scope of civic conversation must equal the scope of the emergency, which is often of national and global dimension. The transition to a regenerative future requires millions—and even billions—of citizens to be in communication with one another. Whatever their points of view, people want to be heard and to have a voice in the future. Diverse communication initiatives begin to provide a vital source of social cohesion in the unraveling world. By the middle of the decade, this recognition ignites a "Community Voice" movement at the local scale and an "Earth Voice" movement at the global scale.

Community Voice initiatives work to mobilize television and to take back the airwaves for a new level of citizen dialogue at the regional scale of major cities around the Earth. An Earth Voice movement works to mobilize the power and reach of the internet that encircles the planet. These initiatives, convened by a diverse community of trusted elders and youth, generally have only two roles: first, to listen to the concerns of citizens and second, to present those concerns for dialogue before the community in the form of "electronic town meetings" and then to "let the chips fall where they may." Successful Community Voice organizations are non-partisan and neutral, and do not advocate a particular perspective; instead, they serve as a vehicle for citizens to have a voice in their own affairs and future. Leadership of one community inspires and catalyzes other communities to create their own Community Voice organizations, and a new layer of robust dialogue begins sweeping through regions and nations. As citizens voice their concerns and electronically vote on different solutions, they begin to break through the gridlock of powerlessness of the past.

By the end of the decade, three quarters of world population will own a mobile phone and have access to the internet. An Earth Voice initiative is underway as people recognize and mobilize the power of the internet as a vehicle for collective attention and action. A majority of Earth citizens realize that, with mobile phones, they literally have in their hands the technology required to engage in a planetary-scale dialogue and to develop a visible consensus for a viable future.

A perfect storm of global crises is growing and challenging humanity to make dramatic changes in how we communicate about how to live on Earth. The human community has entered uncharted territory. Never before have we been so compelled to come together as regions, nations, and the world. The combined power and potential of Community Voice and Earth Voice movements provide practical tools for the unraveling world to weave itself together in new ways.

2030s: The Great Collapse—Free Fall

Summary

The fragile and complex world-system has become so frayed it can no longer hold together and, with unexpected and breathtaking speed, it breaks apart and drops into free fall. Chaos, confusion, and panic sweep through the world. Vital services are interrupted. Police and fire protection become sporadic. Waves of energy blackouts occur as large-scale electrical grids fail. Large institutions (corporations, universities, healthcare systems) go bankrupt, resulting in massive unemployment. Overall, with little to hold the world together, the bottom drops out and we experience the collective panic of a great fall.

Massive indebtedness created by extravagant spending in earlier decades now prevents many institutions from mobilizing resources for creative action. Instead of rising to challenges, many institutions unravel. Bankruptcy spreads to entire cities. Many vital services falter—including police and fire protection, and the upkeep of infrastructure, such as roads and electrical grids. Large corporations go bankrupt, resulting in job losses for huge numbers of people. Major colleges and universities become insolvent and close their doors. Many big churches cannot afford the upkeep and fail. Breakdowns spread in waves throughout the world, and people must increasingly fend for themselves at the local level. Instead of creative action to avert a deepening climate crisis, the world is preoccupied coping with rapidly spreading breakdowns.

Global demand for fresh water increases in excess of availability and roughly three-billion people suffer from water shortages. The diversity of food options declines dramatically as drought reduces agricultural productivity. The number of climate refugees climbs to approximately a hundred-million people migrating to more favorable areas. The civic structures and resources of many nations are overwhelmed completely. Pollinating insects are dying, compromising the world's food supply. The integrity and health of the biosphere (plants, land animals, birds, insects, and ocean life) deteriorates rapidly. Survival pressures are growing so great, little attention is given to repairing and restoring eco-systems.

World population continues to grow, particularly in Africa, approaching a total of nine billion. Divisions and separations of every kind are growing—financial, political, generational, gender, racial and ethnic, and religious. The world is awash with so many disputes at so many levels with so many differences of so many kinds that little room exists for rising to a higher humanity. The world is filled with blame, fault-making, denunciation, hostility, condemnation, and reproach. Challenges to the mindset of consumerism and capitalism grow as millions struggle for their very survival.

An internet-based "Earth Voice" initiative—rich with grassroots dialogue and feedback—takes root in the unraveling world. Media organizations are held accountable for supporting a new level of social communication. As nations weaken, governance is increasingly forced downward to regions, cities, and local communities. Ecovillages, pocket neighborhoods, and other designs for living begin to establish a resilient foundation for sustainable cities. Work roles change dramatically as small, self-organizing communities provide new settings for employment with diverse skill sets suited for localized living. Simplicity is grudgingly accepted as a survivalist approach to living—a way of stopping short of hitting rock bottom.

Review of Major Driving Trends in the 2030s

- **Global Warming and Climate Disruption**: Global temperatures increase over historical levels by 2°C (3.6° F) by the end of the 2030s. With a 2° C increase, ice sheets begin irreversible disintegration that will produce a catastrophic sea-level rise, most dramatically in the next century. In addition to producing warming oceans, shrinking ice sheets and ocean acidification, temperature rise also brings new extremes of storms, rain, floods, and droughts that severely impact agriculture and habitats.[83]

    A 2°C increase is viewed as a critical climate tipping point—the beginning of runaway climate change.[84] The potential for unstoppable warming begins with the release of the "sleeping giant" of methane, roughly 80 times more potent as a greenhouse gas than CO_2.[85] A surge in atmospheric methane threatens to erase the anticipated gains of the Paris Climate Agreement.[86] Furthermore, we face the dire prospect of self-reinforcing feedback loops pushing the climate into chaos before we have time to restructure our energy system.

Another "sleeping giant" is the Amazon rainforest, which has been viewed as a CO_2 "sink" that absorbs carbon. However, a recent study shows that tropical forests are losing their ability to absorb carbon, which will turn the Amazon into a *source* of CO_2 by the 2030s and accelerate climate breakdown, producing much more severe impacts requiring a much faster reduction in carbon-producing activities to counteract the loss of carbon sinks.[87]

- **Climate Refugees**: With climate disruption, the number of refugees climbs from tens of millions on the move to a hundred-million or more migrating to more favorable areas by the end of the 2030s. Migrations of this magnitude overwhelm the capacity of regions to adapt. For perspective, roughly one million refugees destabilized much of Europe in the 2010 decade. With a hundred million or more migrating, the impacts are expected to be many times greater and spread unevenly, mostly throughout the more resource-favored northern hemisphere.

- **Water Scarcity**: Global demand for water exceeds sustainable use by 40 percent.[88] By 2030, at least three-billion people suffer from water shortages.[89] With growing drought conditions, major cities around the world begin to run out of water. In 2019 Cape Town, South Africa, came close to "zero day"—the day when the city runs out of water. Cape Town is just the beginning. At least eleven other major cities are likely to run out of water before the end of the century: São Paulo, Brazil; Bangalore, India; Beijing, China; Cairo, Egypt; Jakarta, Indonesia; Moscow, Russia; Mexico City, Mexico; London, England; Tokyo, Japan; and Miami, USA.[90]

"In India, a country of 1.3 billion people, fully half the population lives in a water crisis. More than 20 cities—Delhi, Bangalore, and Hyderabad among them—will gulp their entire aquifers dry within the next two years. This translates into a hundred-million people living with zero groundwater."[91]

- **Food Scarcity**: For every degree Celsius of temperature rise, a 10 to 15 percent decrease in agricultural yields is expected. Therefore, a 2°C (3.6°F) temperature increase is expected to reduce agricultural productivity by 20 to 30 percent at a time when demand already stretches food supplies to their limits. Pockets of food scarcity grow into areas of outright famine, producing further mass migrations and civic breakdowns. (See the food scarcity list below to explore how diets may be dramatically diminished.)[92]

FOOD SCARCITY

In the coming decades, a range of foods will become prohibitively expensive for all but the most affluent. An illustrative list is below. It is enlightening to go down the list and check off those foods you will greatly miss as they become increasingly costly. Unless you are growing many of these yourself or have considerable wealth, these foods will be virtually unavailable. This is a visceral example of the climate crisis hitting home.

❏ Almonds	❏ Coffee	❏ Potatoes
❏ Apples	❏ Corn	❏ Squash
❏ Avocadoes	❏ Honey	❏ Rice
❏ Bananas	❏ Maple Syrup	❏ Shrimp
❏ Chicken	❏ Oysters	❏ Soybeans
❏ Chocolate (Cacao)	❏ Peaches	❏ Strawberries
❏ Cod	❏ Peanuts	❏ Wine (Grapes)

People begin to create new diets that adapt to reduced options for basic foods. Poorer people are forced to accept diets with diminished nutrition, reduced variety, and less flavor—a significant decline in well-being and quality of life. A food revolution is underway that privileges the wealthy who can buy their way out of food limitations with genetically modified, greenhouse-produced foods at much higher costs.

- **World Population**: Human numbers are expected to reach nearly nine billion by 2037.[93] A world population of nine billion at the end of the 2030s is a realistic estimate, with much of the growth occurring in Africa, India, and southern Asia.

- **Species Extinction**: Building on projections made in the 2020s that estimate a million species could be extinct by the end of the century, the loss of animal and plant species is expected to rapidly accelerate.[94] The integrity and health of the Earth's biosphere (plants, land animals, birds, insects, and ocean life) deteriorates rapidly. Oxygen loss driven by global warming (and nutrient pollution produced by runoffs from agriculture and sewage) suffocates the oceans, with far-reaching and complex biological implications, resulting in a marked decline of ocean life.[95]

- **Economic Growth/Breakdown**: Given the extraordinary demands for making an extremely rapid transition to renewable sources of energy, the global economy is in deep turmoil and crisis. Overall growth comes to a standstill despite dramatic efforts to increase renewable energy. Tremendous economic and social pressures shift more developed nations away from a historical focus on unrestrained economic growth and consumerism.

Around the world, creative experiments are underway to discover practical ways to recreate the economy so it works for both people and the planet. The goal of creating self-organizing, regenerative forms of economic activity that serve global civilization is more widely accepted.[96] With massive displacement of workers through automation—combined with dislocations through climate disruption and the breakdown of large-scale factories and corporations—regenerative approaches to living foster the development of "local living economies."

Regenerative economies nested within alternative forms of community emerge around the world to create more resilient living systems. Nonetheless, an insurmountable magnitude of change seems required to make a global transition to renewable energy and regenerative economies designed with fairness and equity.

- **Economic Inequity**: The richest one percent on the planet are on target to own two-thirds of all wealth by 2030.[97] Vast disparities in wealth, coupled with economic demands for shifting to a net zero-carbon economy by 2050, place extreme pressures on the already disrupted global economy and society. An extreme lack of fairness and trust drains legitimacy from the world's economic system.

With enormous disparities in wealth and incomes, in the 2030s we face a decade of cascading economic breakdowns where vulnerable areas experience full economic collapse. The growth paradigm of materialism and consumerism disintegrates as a compelling social goal — not only does this paradigm undermine the well-being of most people, it also contributes to the devastation of the Earth's biosphere.

Scenario: Imagining How the 2030s Unfold

In the decade of the 2030s, people around the world recognize that a full climate catastrophe is developing. Yet entrenched bureaucracies—for example, in business, media, education, religion, and social services—are still largely unprepared and ill-equipped to meet the challenges of a worsening climate, deteriorating economies, and a collapsing biosphere.

In more affluent countries, most people are deep in debt, taxes are profoundly unequal, and the engines of economic growth are faltering. A rapid turnover of leaders and policy solutions occurs, but nothing seems to work for long. Efforts to create order are overwhelmed by growing levels of disorder. Large-scale social cohesion is alarmingly minimal and many leaders govern with virtually no support.

Previous levels of resilience are depleted in a grinding downward spiral of bureaucratic confusion and chaos.[98] We no longer have the capacity to bounce back quickly from difficulties. Some people seek security by turning toward more controlled, authoritarian districts. Others turn toward self-organizing communities that depend on strong relationships and collaborative approaches to living.

As climate disruption deepens, divisions of every kind increase—financial, political, generational, gender, racial and ethnic, and religious. The only constant of this disorienting and confusing decade is the unrelenting stress produced by breakdowns and separations.

The wealthiest people who enjoy a "good life" of material comfort and advantage face a growing outcry of protest from billions struggling for survival. Nonetheless, wealthy elites resist making rapid adaptations to new ways of living. Having invested their lives and identities in material accumulation, they fight back, claiming their privilege is earned and deserved. Although most recognize the new realities, many reject new norms for living. However, by the end of the 2030s, their efforts to separate into gated and guarded communities begin to falter as billions of impoverished people, with nothing to lose and much to gain, rise in protest.

With breakdowns increasing, localization grows with a rush of social, economic, and technical innovations. Pocket neighborhoods grow into diverse forms of eco-villages, establishing a resilient foundation for transition towns and sustainable cities. Newly organized communities build more than physical structures; they develop a new understanding of human character and a maturity that seeks to serve the well-being of all. Work roles change dramatically, as small self-organizing communities provide new settings for developing diverse skill sets for living.

Pushed by the climate crisis and spreading breakdowns, the affluent majority in developed nations recognize we must transform cultures of consumerism and reduce our ecological footprint if we are to avoid global catastrophe. The cultural hypnosis of consumerism loses potency as people recognize the dream of unrestrained consumption is a devastating nightmare future for the Earth. In reaction, a global culture, valuing simplicity and sustainability, begins to emerge. Mass media advertising that has been aggressively promoting the trance of consumer culture shifts from product commercials to "Earth commercials" as companies proclaim their commitment to a healthy planet.

Wealthy countries are responsible for climate change, but the poor suffer most. Given the disproportionate impact of global warming on poorer countries, wealthier nations are pressed—with only modest success—to take responsibility for supporting climate adaptations. Strong initiatives are vital to foster a sense of global unity and cooperation. Yet climate change is increasingly devastating everyday life in poorer nations—including water availability, food production, health care, environmental quality, and the well-being of vulnerable populations, especially women and children.

In poorer countries, the impacts of global warming often reverse progress in gender equality, as men are forced to migrate to find work, leaving women to handle the entire burden of raising children, farming, or fishing locally, and managing the household. This leaves women more isolated and less able to find meaningful work and education.

In recognizing the adverse impacts of global warming on developing nations, a global movement for compensation, reparation, and adaptation grows, seeking to build a new sense of partnership among the people of the Earth.

The trans-partisan *Community Voice* movements that began in the 2020s now become important sources of social cohesion. These continue to grow around the Earth, connecting humanity into ever-larger communities engaged in intense conversations. Recognizing that the scale of conversation must match the scale of challenges, *Earth Voice* dialogues become firmly established in the unraveling world. Increasingly, people recognize that the mass media are a key component in our "social brain," a direct expression of collective intelligence. The slogan, "As the media go, so goes our future" is widely affirmed. Media organizations are held accountable to an entirely new degree and are mobilized to support humanity's social imagination to visualize pathways of progress toward a sustainable and meaningful future.

Media activism grows into a central force for cohesion as a growing number of institutions break down and break apart. Sorrow and grief increase as loss and tragedy grow around the world. *Collectively witnessed, we realize we are going through this rite of passage together.*

Although the old world is unraveling and local-to-global communication is growing, we still lack the overall support needed to move swiftly into a transforming world. Consumer society and ways of living change slowly, the dispossessed continue to be largely ignored, a green transition is unable to mobilize a majority for dramatic action, and authoritarian districts continue to separate themselves into compartmentalized areas of control. Given deep divisions, the 2030s are a time of churning chaos and conflict, without an overarching set of values and intentions for moving ahead.

Financial institutions go into free fall. Local and national governments, financial organizations, academic institutions, religious organizations, to name a few, are overwhelmed trying to understand what is happening and are significantly under-resourced when trying to respond. Yet the struggle for a new paradigm of living is getting underway. People ask: *How can we once again feel at home on the Earth?* Do we have the collective maturity to consciously make a great transition to a new future?

2040s: The Great Initiation—Sorrow

Summary

In the decade of the 2040s, most people recognize that we are losing the race with climate catastrophe. Runaway climate disruption is no longer only a looming possibility—it is an overwhelming and vividly present reality. As the consequences of climate chaos, financial breakdowns, civic anarchy, species extinction, mass migrations, and widespread famines continue to grow, the entire world moves toward unstoppable collapse. The need for a profound transformation is anchored in the raw experience of humanity. We recognize that we either pull together in common effort or face the functional extinction of our species. We understand that Earth will never go back to the climate patterns of the previous 10,000 years since the end of the last ice age. We accept feelings of shame, guilt, grief, and despair as a ruinous future grows around us.

The biosphere is increasingly impoverished, weakened, and barren. Profound climate disruption, falling agricultural productivity, extreme water scarcity, and great economic inequities create huge areas of devastating famine. This is also a time of "great burning," as unrelenting droughts dry out the land and fire scorches vast regions of the Earth. And it is a time of "great dying," too, as millions of people and countless species of animals and plants perish. Humanity confronts a two-fold tragedy of unimaginable proportions that shocks and awakens the soul of our species.

Broken supply chains lead to hoarding, looting, black markets, and hyperinflation. Adaptations are pushed down to the local level of neighborhood and community, and people search for others they can trust and work with in rebuilding life from the ground up. Old sources of value (measured in cash, stocks, and bonds) have become nearly worthless. New sources of value reside in strong personal relationships and access to scarce resources such as food, medicine, and fuel that have tangible importance. Despite its great value, an *Earth Voice* movement struggles to stay alive as the internet is constantly breaking down and being repaired.

The world descends into collective despair. Feeling we have failed to step up to our responsibilities as planetary citizens, many mourn for the lost Earth. The soul of humanity is grievously wounded with moral injury. We face a future of unending bleakness and despair—unless we rise collectively to this time of challenge.

Review of the Major Driving Trends of the 2040s

- **Global Warming and Climate Disruption**: In this decade, we move beyond 2°C (3.6°F) of warming toward a new benchmark of 3°C (5.4°F)—a critical climate-tipping point.[99] Methane surges into the atmosphere, triggering runaway feedback loops.[100] The world moves beyond breakdowns, toward full collapse and climate catastrophe. An already turbulent and chaotic climate grows to catastrophic proportions. Climate extremes include both fire and water—large regions of the Earth experience unprecedented drought that brings fires to a scorched Earth, while other regions experience unprecedented storms, floods, and sea-level rise.[101]

- **Water Scarcity**: Water shortages are critical for three billion (or more) people. In turn, water scarcity produces a dramatic increase in the number of climate refugees fleeing drought-stricken regions.
- **Food Scarcity**: Growing population pressures combined with climate disruption, falling agricultural productivity, water scarcity, and economic inequities to produce large areas of devastating famine.
- **Climate Refugees**: At least 200 million climate refugees are expected to be on the move, creating colossal, social, and economic disruptions, as communities in resource-favored areas try to cope with the influx of overwhelming numbers of people.
- **World Population**: In the 2040s, population continues to grow and runs up against increasingly severe limits created by water and food scarcity and the breakdown of ecosystems.[102] Tragically, it seems plausible that ten percent or more of the Earth's poorest and most vulnerable populations will be at great risk of dying during this period of great transition. With a global population of roughly nine billion people in the 2040s, this means that roughly 900 million people could perish. These millions will not die quietly and out of sight, but in our media-rich world, will die very publicly, painfully, and visibly. Their deaths will be caused by famine and disease, as well as by enormous levels of violence in conflicts over dwindling resources.

  The die-off of hundreds of millions of people will produce unimaginable levels of moral and psychological trauma. The needless suffering and death of hundreds of millions awakens humanity to choose a path of greater equality and fairness in how we live together.
- **Species Extinction**: Decades of ecosystem destruction undermine the foundations for life globally. Countless species go extinct, leaving the Earth an ever-more barren world. The unyielding reality of ecological collapse confirms that we are an integral part of the global web of life and that the threat of extinction applies to humans, as well.

- **Economic Growth/Breakdown**: Economic breakdowns spread around the world, producing full-scale collapse of vulnerable economies. Although economic breakdowns slow greenhouse gas emissions, widespread efforts for survival have the unfortunate result of pushing people and communities to use whatever energy sources are readily available, including coal and oil, for short-term survival. A return to fossil fuels contributes to greenhouse gas emissions at the very time we must reduce them. Although efforts for deep reconfiguration of the local-to-global economy are underway, collapsing economies and eco-systems make these efforts exceptionally difficult.

- **Economic Inequities**: The immensely complex and difficult transition to a global economy operating on renewable energy reduces overall output, and civilization is more challenged than ever to meet the needs of the world's poor and to move toward much greater fairness. Global tensions between the haves and have-nots accelerate past breaking points. The global crisis of fairness and social justice conflicts with cultures of consumerism, resulting in a fierce struggle for the future direction of our species.

People with the least access to resources face the greatest challenges in adapting to global warming—and this is true across race, gender, age, geography, and class differences.[103] Widespread efforts grow for producing the essentials of life inexpensively and to restrain luxury lifestyles for the affluent. The redistribution of land is also a key factor in fairness and awakens titanic struggles over ownership and sharing.

Scenario: Imagining How the 2040s Unfold

In this decade, we move into a time of great suffering beyond anything humans have ever experienced.[104] A global collapse is underway, generating all kinds of shortages—including vital medicines and medical care, basic foods, and clean water. Many major corporations go bankrupt as their consumer base disintegrates. Major cities also go bankrupt as their tax base dissolves. Key infrastructure is abandoned and falls into disrepair as nearly all maintenance is neglected—electrical and telephone utilities, internet services, roads, bridges, traffic lights, sewage systems, garbage removal, and water systems.

Confusion, chaos, and conflict grow. As lawlessness spreads, private protection forces replace traditional police and law enforcement. At a larger scale, collapse spreads beyond cities to states and even nations. As nations descend into bankruptcy and break apart, so do international organizations such as the United Nations, which endure as little more than symbolic entities. Global cohesion is sustained and shaped not by international institutions, but by a fast-growing, electronic commons emerging from grass-roots movements around the world. These grassroots movements use the faltering global communications infrastructure to create a new global commons in our collective consciousness.

Neither the public nor private sector has the resources to mount large-scale projects that might offer a meaningful response to the magnitude of collapse underway. Adaptations are pushed down to the local level of neighborhood and community, where people must rely on the people, skills, and resources available nearby.

In the 2040s, much of humanity's story can be told under two headings: the "Great Dying" and the "Great Burning." Although tens of millions of people perished in the previous decade, human die-off escalates, and a horrific period of "Great Dying" begins in the 2040s. The carrying capacity of the Earth is estimated to be roughly three-billion people living middle-class European lifestyles. A global population approaching nine-billion is far beyond the Earth's estimated carrying capacity.[105] Humans discover that we are no different from the rest of life on Earth facing extinction.[106] A tidal wave of death sweeps across the planet, bringing unrelenting disease, famine, and violence that stain the soul of our species.[107]

The mathematics of death are unrelenting. With roughly nine-billion people on the planet in the 2040s and, conservatively, with ten percent of the world's population (the poorest of the poor) at greatest risk of dying, it means that 900,000,000 people could die in this ten-year period. Basic arithmetic translates this figure into an astonishing 90,000,000 people dying *each year*—roughly the equivalent of seven holocausts for *each year* in this decade.

As waves of death sweep the Earth, the moral and psychological impact of these losses staggers the human psyche. This calamity unfolds in real time with high-definition media revealing the faces and passing lives of countless humans and other creatures. The immeasurable pain and suffering of the Great Dying tears apart the fabric of culture and consciousness. The loss, grief, and sorrow are incalculable. These wrenching years shred our connections with the past and leave our legacy in tatters.

The magnitude of tragedy and suffering in the Great Dying transforms the heart and soul of our species.[108]

The second area of great tragedy and suffering marking this decade is the "Great Burning."[109] Although extreme fires have been burning in localized areas around the world since the 2020s, raging fires throughout the planet become a dire emergency two decades later. As global warming intensifies, areas of severe drought and great burning also intensify.

- Much of the Amazon has dried out and is burning.[110]
- Large swaths of California and the western United States are chronically on fire, transforming ancient forests into scrubland and brush.[111]
- Large areas in the Los Angeles region burn, as do large regions in Texas and Colorado.
- Considerable portions of Mexico are aflame.
- Much of Australia is incinerated.[112]
- Large regions of Europe—especially southern France, Portugal, and the rest of the Mediterranean region—are burning.
- Major portions of India, Pakistan, Iran, and Afghanistan are on fire.
- Regions of northern and southwestern China are regularly in flames.
- Large areas in Africa are chronically ablaze—especially Ethiopia, Uganda, Sudan, and Eritrea.

Instead of labelling our age as the "Anthropocene," in his book *Fire Age,* Professor Stephen Pyne defines it as the "Pyrocene"—a future with fire and upheavals so immense and unimaginable that "the arc of inherited knowledge that joins us to the past has broken" and we move into a future unlike anything we have known before.[113]

The "Great Burning" and the "Great Dying" symbolize the functional disintegration and disconnection of human civilizations with the past. We are literally no longer able to function as we did before. Despite the great efforts of previous decades, humanity's evolutionary experiment is failing. The last vestiges of trust in humanity's historical path of material progress are drained from the world.

The powerful elites who dominated the globe in prior decades retreat into enclaves as the world falls apart around us all. The planetary eco-crisis accomplishes what nonviolent action and protest could not—the *awakening of humanity*. Above all, humanity needs a new and purposeful way forward, as well as a strong vision and voice for getting us there.

The human population collectively experiences CTPS (Chronic Traumatic Planetary Stress) an entirely new mindset encompassing the entire human family. The difference between PTSD (post-traumatic stress disorder) and CTPS is that instead of a relatively brief and confined episode, the trauma becomes life-long and planetary in scope. There is no escape—the burden of collective trauma permeates the soul of humanity.

Even while absorbing this decade of immense suffering, people realize that our deteriorating biosphere will produce still greater suffering in the coming decades as people cope with being torn from their roots of land, culture, community, and livelihood. Although this has occurred in the past, it becomes a planetary-scale phenomenon in the 2040s. Consequences of CTPS include:

- Extremely high levels of social anxiety, fear, and protective responses,
- A narrowing focus of attention and difficulty concentrating on the big picture,
- Emotional numbing and widespread use of alcohol, drugs, and media for escapism
- Reactivity, violence, and mood disorders,

- Feelings of helplessness, hopelessness, and depression leading to epidemics of suicide.

The incalculable suffering of this decade dissolves old identities and dogmas, leaving many deeply wounded, both psychologically and socially. Stress expert Hans Seyle wrote: "Every stress leaves an indelible scar, and the organism pays for its survival after a stressful situation by becoming a little older."[114] At the very time we need to pull together in cooperation as a species, CTPS makes it much more difficult.

The immense suffering of these times is not without merit. In the consumerist pursuit of ongoing happiness, many lost touch with the depths of life—with our souls. For more than two decades, psychotherapist Francis Weller has worked with groups, facilitating authentic encounters with grief. Weller writes:

> "To traditional people, soul loss was, without doubt, the most dangerous condition a human being could face. It compromises our vital energy, decreases joy and passion, diminishes our aliveness and our capacity for wonder and awe, saps our voice and courage, and ultimately, erodes our desire to live. We become disenchanted and despondent."[115]

A great gift lies hidden in great grief—a passageway to reconnect with our soul. Carl Jung advised, "Embrace your grief, for there your soul will grow." Unacknowledged sorrows limit contact with the collective soul of our species. As humanity encounters the darkness of our collective losses, we recover contact with our communal soul. Francis Weller writes:

> ". . . without familiarity with sorrow, we do not mature as men and women. It is the broken heart, the part that knows sorrow, that is capable of genuine love. . . . Without this awareness . . . we remain caught in the adolescent strategies of avoidance and heroic striving."[116]

Grief challenges the unspoken agreement of consumer society to accept lives that are shallow and unfeeling. Grief is an entrance to the natural undomesticated aliveness of our soul. Welcoming grief is the secret to being fully alive—the doorway to the wild untamed vitality of the soul. Naomi Shihab Nye, in her poem "Kindness" writes:

Before you know kindness as the deepest thing inside,
You must know sorrow as the other deepest thing.
You must wake up with sorrow,
You must speak it till your voice
Catches the thread of all sorrows,
And you see the size of the cloth.[117]

The magnitude of the world's sorrow is immense. We discover what the indigenous soul has always known: *We are not separate from the Earth—aliveness is everywhere and in all things.* When the Earth is impoverished, we are impoverished in equal proportion.

Humanity has so much to grieve because the losses are so great: In the Great Dying, we lose millions of precious fellow beings—sisters and brothers seeking their unique life on Earth, their potentials unrealized, relationships unfulfilled, talents unexpressed, gifts unreceived by others. We also lose so much of the rest of life—the plants and animals that bring richness, resilience, and beauty into our lives.

In the 2040s, we not only lose countless lives, we also lose cities, cultures, languages, and wisdom. For example, with sea-level rise, we lose many of the world's oldest cities established on the seacoasts—Alexandria, Egypt; Shanghai and Hong Kong, China; Jakarta, Indonesia; Mumbai, India; Ho Chi Minh City, Vietnam; Osaka and Tokyo, Japan; London, England; New York and Washington, DC, USA . . . and many more.[118]

Losses are so widespread and so foundational that they awaken people to the wisdom of *ubuntu*: "I am who I am because of who we are." When the feeling of "we" is diminished, I am diminished in proportion to the richness of life that has been lost. When we are in contact with our essence, our soul, we are immersed in the larger ecology of aliveness. We share in the kinship of all beings and experience directly the subtle hum and song of all life on the planet.

In the grip of overwhelming sorrow for the immensity of our losses, we yearn to return to where we were before grief overtook us. Yet we know we can never go back; instead, we are challenged to accept our fate and discover how this wisdom can transform our path into the future. Collective grief burns through fabrications and facades, and we encounter our raw humanity. In the authenticity of that encounter, we move forward to build new worlds.

> *In the sorrow of the Great Dying and Great Burning,*
> *we are naked to evolution. Sorrow is no con game.*
> *This is the real world.*

When sorrow claims us, we know this world is not make-believe. We face the honesty of life itself, to be honored and accepted for what it is. Jennifer Welwood, a teacher of spiritual psychology and a poet, speaks to these times:

> *My friends, let's grow up.*
> *Let's stop pretending we don't know the deal here.*
> *Or if we truly haven't noticed, let's wake up and notice.*
> *Look: Everything that can be lost, will be lost.*
> *It's simple—how could we have missed it for so long?*
> *Let's grieve our losses fully, like ripe human beings,*
> *But please, let's not be so shocked by them.*
> *Let's not act so betrayed,*
> *As though life had broken her secret promise to us.*
> *Impermanence is life's only promise to us,*
> *And she keeps it with ruthless impeccability.*
> *To a child she seems cruel, but she is only wild,*
> *And her compassion exquisitely precise:*
> *Brilliantly penetrating, luminous with truth,*
> *She strips away the unreal to show us the real.*
> *This is the true ride—let's give ourselves to it!*
> *Let's stop making deals for a safe passage:*
> *There isn't one anyway, and the cost is too high.*
> *We are not children anymore.*
> *The true human adult gives everything for what cannot be lost.*
> *Let's dance the wild dance of no hope!* [119]

Grief takes us beyond hope to the raw truth of reality. In our collective grief, we are called to move beyond our species' adolescence, to acknowledge our actual situation, to show up for what is real and to respond as best we can.

The Great Dying cries out for our collective maturity beyond hope or despair—and summons us to step up and simply take responsibility for doing the work called for by our times of Great Transition.

Grief reveals the depths. In the encounter with death, we are ready to turn more fully to life. As we encounter what seems most unbearable, we discover what is most poignantly alive. Grief demolishes pretense and cuts through the superficial happy talk of consumer culture. We have reached a hinge-point in history where humanity must make choices with consequences cascading into the far future. This is evolution in the raw. The Great Dying calls us to a higher level of collective maturity—to reach beyond our species' adolescence to take charge of our future.

Collectively, we wonder whether we have the maturity to place the well-being of life above our personal interests. Can we engage these difficult times with humility and compassion? Can we talk less and listen more to the suffering of the world? Can we take charge of how we live and work to create a habitable biosphere, understanding that this requires a dramatic change in our manner of living?

Particularly in wealthier nations, a deep psychological crisis has developed as people feel enormous guilt and shame for the devastation of the planet and the diminished opportunities for future generations. Many are in mourning for the Earth and feel that humanity has failed in its grand experiment in evolution. After tens of thousands of years of slow development, many feel that within the span of a single generation we have ruined our chance at evolutionary success and grieve this lost opportunity. The human community recognizes that we face a bleak future of widening ruin and deepening despair unless we collectively rise to this time of challenge.

Suffering and sorrow are a cleansing fire that awakens the soul of our species. Waves of ecological calamity have reinforced periods of economic crisis, and both have been amplified by massive waves of civil unrest. Momentary reconciliation is followed by disintegration and then new reconciliation. In giving birth to a more conscious, sustainable species-civilization, humanity moves back and forth through cycles of contraction and relaxation, until we utterly exhaust ourselves and burn through the remaining barriers that separate us from our wholeness as a human family.

Finally, we know with unyielding certainty:
We have a choice between extinction and transformation.

In the 2040s, many wonder whether humanity's demise would be a tragedy or a blessing.[120] Are we such a precious contribution to the Earth that we deserve to live, while a million other species do not? A deep moral crisis pervades the Earth. Are we worthy of continued existence? Can we find a path and purpose that enables us to rise above these tragedies and be worthy of life?

Efforts at reconciliation begin with a sense of promise and hope, only to fall back in the face of climate chaos and systems breakdown. Is there truly a basis for living together on this small Earth with so many differences? We know that the sorrows and divisions of a broken world must be accepted before they can be healed—acknowledging our brokenness is the first step on a path toward wholeness.

Pushed by dire necessity, innovations in building new kinds of community appear. People retrofit old structures to create new expressions of community, ranging from pocket neighborhoods and co-housing to eco-villages of diverse designs. Life-boat communities proliferate as people recognize that smaller-scale constructions can adapt rapidly to changing circumstances. Realizing the importance of healthy communities, support for transition towns and sustainable cities grows, but injury to the earlier economy, society, and ecology is so great that this is immensely difficult. Simultaneously, tensions rise as waves of climate refugees search for security and survival and seek to move into healthy communities.

No longer is simplicity of living regarded as a regressive way of life. Low-carbon lifestyles and accompanying values bring a new regard for community, sufficiency, and kindness. Simplicity of living encourages strong communities of mutual support and survival. As people develop a range of skills that directly contribute to the well-being of their neighbors, they feel their true gifts are welcomed into everyday life.

Forces of uplift are present throughout the world, but they are so fragmented and disconnected from one another they cannot converge into powerful updrafts of mutually reinforcing feedback. The world is broken. Eco-collapse brings ego-collapse. The collective psyche of humanity is desperately wounded. Calls for maturity grow, only to be overwhelmed by forces of disintegration that push humanity down to primal levels of struggling to live. Small communities become the basic scale of security and survival.

Reflective- or witnessing-consciousness grows as humanity is driven to look deep beneath everyday life and to recognize the wounded existence we have created as the foundation for our future. We recognize these transitional times will likely result in either a final descent into functional extinction or arising together in awakening and rebuilding.

Collective communication seems to offer the greatest potential for rapid renewal. Communicate or perish! As we confront the reality of profound ecosystems collapse, we know we cannot retreat from public dialogues and consensus building; yet, for many, local-to-global communication aimed at discovering a way forward seems fruitless and destined for failure.

2050s: The Great Transition— Early Adulthood

Summary

The Great Dying and the Great Burning leave no doubt that the world of the past is gone. Humanity can descend into the darkness of authoritarianism or the ink-black of extinction—or choose to move forward from the deep sorrow in our collective soul to a future of unexpected aliveness. Our time of collective choice is unyielding and urgent. We know intimately the words of poet Wallace Stevens:

> *After the final no there comes a yes*
> *And on that yes the future world depends.*[121]

What will be the "yes" of humanity? "Yes, we surrender" to either authoritarianism or functional extinction. Or: "Yes, we make a courageous choice" to step up to a higher maturity and a transforming future!

As the reality of an unfolding climate catastrophe and whole-systems crisis hits home, the human community is pushed back upon itself to authentically reconsider how to move forward. Can we transform how we collectively think (our species mind) and how we view our purpose for living on Earth (our species journey)? The last three decades brought shattering despair and grief. We have given up the project of trying to reclaim the past. Can we build a new future by awakening a new sense of our species journey? Do we have the social will to make this great turning? Joanna Macy sums up the situation clearly:

> "[Are we]... serving as deathbed attendants to a dying world or as midwives to the next stage of human evolution? We simply don't know. So, what is it going to be? With nothing to lose, what could hold us back from being the most courageous, the most innovative, the most warm-hearted version of ourselves that we can possibly be?"[122]

The deep wounds inflicted by the "Great Dying" and the "Great Burning" haunt humanity's collective psyche. We have been liberated from the shallow trance of materialism and can return to our original intuition of aliveness permeating the world. The paradigm of aliveness honors the spiritual roots of all the world's great wisdom traditions and brings a healing perspective into the world. Initiatives for broad and deep reconciliation can grow and spread from this foundation, and can begin to heal our many divisions—racial, ethnic, religious, wealth, and gender.

At the opening of the decade of the 2020s, we recognized that building a habitable Earth would require rapid reduction of CO_2 emissions to net-zero emissions by 2050. Now this decade arrives with the frightening awareness that humanity's efforts, although heroic, were far too little and much too late. We have not reached this critical goal.[123] Multiple tipping points have been crossed, methane continues to pour into the atmosphere, and global temperatures edge toward a terrifying 3°C increase, producing climate extremes disruptive to all forms of life. A billion people have become climate refugees.

Step by step, we begin to move forward into our early adulthood as a species. With profound regard for the well-being of all life as the foundation for the future, we feel the heartening breeze of uplift and emergent possibility. *Community Voice* initiatives sprout at the regional level and a robust *Earth Voice* initiative is blossoming globally. We know, in our bones, that we are all citizens of the Earth and we seek new dialogues to integrate this understanding into our everyday lives, and to rise together in rebuilding the Earth as our welcoming home. The immense sorrows of the past decade awaken a collective commitment for creating a path into the future that moves beyond the endless distractions of violence.

Recognizing the urgency of finding a higher common ground of understanding and healing, the world immerses itself in an ocean of communication. Around the clock, a rich and complex global conversation searches for understanding and a healing vision for the future. We have crossed the threshold into a new stage of adulthood where we are willing to work for the well-being of all life and make commitments for the deep future. Millennia of work lie ahead as we reconcile ourselves with living together and building a thriving future on a severely wounded Earth.

Review of Major Driving Trends in the 2050s

- **Global Warming and Climate Disruption**: The goal of zero CO_2 emissions by 2050 is not reached. Global temperatures increase toward a terrifying 3°C (5.4°F) and produce highly disruptive and destructive climate changes.[124] Methane continues to pour into the atmosphere, amplifying extreme weather patterns, reducing agricultural productivity, impacting coastal areas with storm surges and hurricanes, and profoundly disrupting the habitats for plants and animals. With unprecedented warming and acidification, the oceans are largely depleted of life, the soil is cooked and dry, and ecological breakdowns are widespread as plants and animals cannot adapt to the speed of climate change.

For several decades, we have recognized that, if temperature increases to 3°C, the chance of avoiding four degrees of warming is poor and, if we get to 4°C, it will produce even more intense feedback loops that will make it extremely difficult to stop temperature rise at 5°C.[125] We are on a roller-coaster ride to hell.

The full climate crisis has arrived.

- **Water Scarcity**: Water stress is expected to impact 52 per cent of the world's population by 2050."[126] With world population approaching ten billion, this means more than five billion people will likely suffer water scarcity.[127] (This estimate ignores the likelihood of a period of Great Dying where a billion or more humans perish.) For many, living has become a miserable struggle for survival in an over-heated and parched world.

- **Food Scarcity**: By 2050, global population is expected to grow beyond nine billion, yet food supplies are under enormous stress and are imperiled as the world moves toward an increasingly barren ecosystem that lacks a rich diversity of plant and animal life. Food demand is 60 percent higher than in 2020, yet global warming, urbanization, and soil degradation have shrunk the availability of arable land.[128] Recall that each degree Celsius (1.8 Fahrenheit) of warming is estimated to produce a 10 to 15 percent decrease in agricultural yields. Therefore, at 3°C, agricultural productivity falls by 30 to 45 percent as a result of rising temperatures. Compounding the situation, efforts to reduce carbon emissions include reducing the use of petroleum-based fertilizers and pesticides. Unable to prop up agricultural production, food stocks fall further, and billions of people are at risk of starvation. "As many as five billion people . . . face hunger and a lack of clean water by 2050, as the warming climate disrupts pollination, freshwater, and coastal habitats. People living in South Asia and Africa will bear the worst of it."[129]

- **Climate Refugees**: Upwards of 300 million climate refugees are expected by mid-century and could be much higher.[130] The flood of an enormous number of refugees into more habitable regions of the planet sets the stage for enormous conflicts.

- **World Population**: The world has grown to an estimated ten billion people by 2057.[131] However, this estimate does not factor in the "Great Dying" of the 2040s, where ten percent or more of the world's population could perish. The potential magnitude of death in the 2050s seems unimaginable, especially with increasing water scarcity and declining agricultural productivity.

- **Species Extinction**: Habitats for plants and animals around the Earth—on land, in the oceans, and in the air—are profoundly disrupted at speeds far in excess of their ability to adapt. By mid-century, roughly one-third of all life on the planet is dying, with horrendous results. The death of entire species of insects leads to a cascading collapse of the biosphere. The quantity and character of food supplies are dramatically altered. Grasslands are imperiled. Animals that rely on plants for food are imperiled. The short-term beneficiaries of these die-offs are scavengers—cockroaches and vultures on land and jellyfish in the oceans.[132]

- **Economic Growth/Breakdown**: By mid-century, the impacts of global warming are dire. Efforts to cut carbon emissions to zero depress economic growth and are regarded as a failure—compounded by a growing wave of economic breakdowns, bankruptcies, and organizational disintegrations. Shortages of all kinds increase, accompanied by hoarding, black-markets, widespread theft, and violence. Traditional sources of value (cash, stocks, and bonds) continue to decline while the value of scarce medicine, foods, and fuels rise. Agricultural productivity continues to drop as temperatures rise. Climate disruptions and massive human migrations profoundly disrupt patterns of trade and production. The global economy fractures and fragments, transitioning to local living economies. The growth mindset of the past is largely replaced by a survival and sustainability mindset, with an emphasis on building local resilience in living economies.

- **Economic Inequities**: Extreme inequities persist despite attempts to create a fairness revolution. The impact of global warming is felt most intensely by people least responsible for creating it and least able to mitigate it. The world's poor cope with famine, disease, and dislocation. Extreme poverty—without access to essential tools and resources required for building a viable local economy—forces people into survival living and prevents them from joining in efforts for building an eco-civilization for the Earth. Greater fairness in access to basic technologies and resources is essential to improve the health and productivity of the disadvantaged and to create the foundation for a more sustainable future. Improving the living circumstances of those most impoverished is more than an expression of compassion, it is the way to mobilize a high-leverage, grassroots response to climate disruption and global breakdowns.

Scenario: Imagining How the 2050s Unfold

The Great Dying continues as millions of people perish each month. The shadow of needless suffering darkens the world and permeates humanity's outlook on the future. The "Great Burning" is accelerating, as runaway global warming picks up speed. Millions of climate refugees seek to move into resource-favored areas. Well-intentioned efforts by local communities to share resources are met with massive waves of refugees that rapidly overwhelm already overstretched systems. Many communities find themselves challenged far beyond their capacities. Overwhelm leads to violent conflicts as people and communities are pressed to the edges of survival. Violence fosters local isolationism and a "wall-building" mentality.

Particularly in developed nations, a deep psychological crisis continues to grow as people see diminished opportunities for future generations. Many sink into deep despair. The soul of humanity is grievously wounded with moral injury—we have devastated the Earth and violated our intuitive sense of ethics. We face a future of unending bleakness. Do we have the social will to make a great transition?

The question of questions is:
How can the human community rise together
to meet, with solidarity, the challenges we now face?

We face an existential crisis as a species and are forced to ask, again and again: Who are we? Where are we going? We are pushed to recall the original wisdom that we live within a world permeated by subtle aliveness. Reclaiming the wisdom of deep aliveness connects us with the universe as a unified whole. Our sense of identity and evolutionary journey are being transformed. Increasingly, we regard ourselves as both biological and cosmic beings learning to live within an ecology of aliveness. In breaking the consumer trance of shallow materialism in a dead universe, we are liberated to explore ways of living in a sentient universe that offer great depth of meaning and purpose.

Pushed forward by immense loss, and pulled ahead by the promise of a healing journey, the global nervous system awakens with a new capacity for collective self-awareness. A new "species mindedness" or Earth-scale reflective consciousness emerges. We have begun to develop the capacity to observe ourselves—to know ourselves in the mirror of our collective mind—and to guide ourselves to higher levels of organization, coherence, and connection. With reflective consciousness, we can more clearly witness what is unfolding in the world and more consciously choose our pathway ahead. We move outside the bubble of distracted materialism into wakeful participation with life.

The human family now recognizes that it was our ability to communicate that enabled us to evolve over thousands of years to the edge of planetary civilization. We further recognize how we need a new level of planetary communication that enables us to collaborate and work together for the well-being of all. By the 2050s, we are three generations into the global communications revolution, and have a strong aversion to being manipulated by consumerist media. We recognize our survival depends on an accurate and realistic understanding of what is happening in the world, and we have grown very distrustful of any attempt to have our collective mind manipulated for power and profit. We have a species memory of being flooded with deliberate distortions and misinformation to create chaos, confusion, and distraction.[133] These painful experiences serve as social immunization to reduce the possibility of infections in our collective mind.

A key development for global consensus-building is the rise of supercomputers that have such enormous capacities they can easily monitor the voting of billions of people in real time. By combining the power of artificial intelligence with the trusted records of blockchain technologies, super-computer systems can ensure confidential voting by billions in secure networks. With these advances, the Earth is energized by new levels of local-to-global communication. *Community Voice* organizations proliferate locally and a robust *Earth Voice* organization functions globally. The world is infused with clear communication about our common future. Most people welcome a growing sense of:

- *Identity* as Earth citizens. A global-scale identity does not diminish other identities of nationality, community, ethnicity, and so on; rather it acknowledges the reality of interdependence and responsibility of all citizens for the well-being of the Earth.
- *Empowerment* as Earth citizens. Decades of participation in diverse electronic forums have demonstrated that citizen-feedback can have a powerful influence on public policy.
- *Equality* as Earth citizens. Despite differences of wealth and privilege, in electronic forums, every person's voice and vote count equally in choosing humanity's future.
- *Solidarity* as Earth citizens. Decades of trauma and suffering have created new bonds of trust and recognition that securing a transforming future will be a team effort.

A promising path to a regenerative, purposeful, and sustainable future emerges. Although we have gone to the very edge of ruin as a species, with local-to-global dialogues and with new levels of collective maturity and insight, we have pulled back from the brink of disaster. After exhausting all hope of partial solutions, we have begun to reach beneath the chaos and sorrow of these times and discover a deeper sense of community and collective purpose. We have been living through a Great Dying and are now maturing into a Great Awakening as an Earth community. We are moving beyond self-centered adolescence as a species and into early adulthood with growing concern for the well-being of all life. Recognizing structural racism, extreme inequities of wealth and well-being, gender divisions, etc., we search for healing and a higher common ground that embodies a new level of cooperation and collaboration.

The world is now in a race between extinction and transformation. The collapse of civilizations has not yet irreparably damaged the foundations for building a workable future for the Earth. New configurations of living emerge around the globe—oriented toward small-scale, self-organizing, and self-sufficient eco-villages.

Voluntary simplicity becomes a core value and touches everything—the food we eat, the work we do, the homes and communities in which we live, and far more. Ecological lifeways blossom with a multitude of expressions. People recognize that the restoration and renewal of Earth as a habitable life-support system will take centuries to accomplish, but that journey is now underway.

An array of uplifting factors has emerged from the dark night of the species soul to generate a strong commitment to building a new world. When these seven factors come into play and begin to mutually reinforce one another, they collectively create a lift strong enough for humanity to rise above the downward pull of either authoritarianism or extinction. We recognize we have been moving through a propound initiation as a species and a future of restoration and renewal is possible if we will choose our path ahead consciously. Half-hearted choices will not suffice. Evolutionary advance requires the full commitment of humans to save the Earth and our own future.

2060s: The Great Freedom— Choosing Earth

Summary

A majority of humans recognize we are moving through a choice-point in history. The nurturing Earth that supported our rise to the edge of a global civilization has been transformed by fires, floods, droughts, famines, diseases, conflicts, and extinctions. Rather than put these challenges behind us, we know our work is to accept and integrate them within us. Acceptance is the source of foundational learning, enabling us to endure into the distant future.

The journey of transformation calls us to our maturity, and to establish ourselves as a dynamically stable, self-referencing, and self-organizing species. A new economy begins to grow around the world. Ecovillages and larger communities become engines of a new kind of commerce as they engage with other communities, using local currencies to exchange skills and services—such as education, healthcare, elder care, solar power and wind power systems, organic gardening, hydroponics, vertical farming, home building skills, among others. Resilient ecovillages aggregate into resilient communities and these aggregate into resilient regions of cooperative living.

Increasingly, reverence and care for the well-being of life is grounded in an emerging understanding that the universe is itself a vast living organism of which we are an integral part. We are more than biological beings, we are "bio-cosmic" beings learning to feel at home in a living universe. No longer is reflective consciousness regarded as a spiritual luxury for the few; now it is seen as an evolutionary necessity for the many.

A majority of people consciously choose to work on behalf of an Earth community founded on freedom, equality, ecological well-being, simplicity of living, healing and restoration of the planet, and authentic communication. A vibrant *Earth Voice* movement offers growing coherence and direction to this species intention.

An Earth-sized species-organism, composed of billions of individuals, awakens as collective humanity. With growing solidarity, we choose Earth as our enduring home. At the cost of unspeakable suffering and sorrow, we have burned through the isolations of the past to discover a deep, soulful relationship with the Earth and other humans. We feel we have paid our dues—the price of admission into the first stage of global maturity—through our immense suffering. Great anxiety as to whether our species would survive is replaced by intense feelings of global community, solidarity, and kinship—generating new waves of optimism. We made it through this time of profound initiation *together*. Our species moved through the time of greatest danger imaginable, and we survived. We have truly begun to know ourselves as a human family with all of our faults and idiosyncrasies. We know there is no final rest—that we must work forever to reconcile ourselves with ourselves—and we now know we are equal to the challenge.

Review of Driving Trends for 2060s

- **Global warming** rapidly approaches a catastrophic level of 3°C (5+°F) and the world's climate turns chaotic. Pushed by dire necessity, the world begins to turn toward the large-scale use of climate geoengineering to limit global warming. "Solar geoengineering" to reflect a small proportion of the Sun's energy back into space helps to suppress the temperature rise caused by increased levels of greenhouse gases. A thin shroud of particles fights global warming by imitating the fine ash from volcanic eruptions that deflects solar radiation streaming into the atmosphere. While this shroud of particles offsets the rapid rise in global temperatures, the reduction of solar radiation is also expected to produce dramatic changes in weather systems and rainfall patterns driven by solar energy. For example, with solar geoengineering, the Asian monsoons, on which two billion people depend for their food crops, could begin shutting down. Despite enormous risks, solar geoengineering is likely to be implemented on a planetary scale by the 2060s in an effort to stabilize global warming. Global warming could also be mitigated with massive efforts for carbon capture that include, for example, planting a trillion or more trees around the planet.

- **Water scarcity** stresses more than half of the world's population, generating intense conflicts and violence over access to water. A planetary-scale initiative gets underway to allocate access to water and to develop desalination plants powered by solar energy.

- **Food scarcity** grows as populations increase and productivity falls. Half of the world's population confronts chronic scarcity and famine. As with water, a global initiative to ration and allocate food also gets underway.

- **Climate refugee** numbers continue to grow dramatically. Cornell University estimates that by 2060, an astonishing 1.4 billion people—about one-fifth of the world's population—could become climate-change refugees.[134] The civic structures of an unraveling world will be overwhelmed and require global cooperation to find suitable homes for people.

- **Species extinction** accelerates as plants and animals fail to adapt swiftly enough to the dramatic changes in climate and weather patterns. As biosphere degrades, a growing fraction of humanity could sign up for a volunteer corps to work for renewal of the Earth.
- **Economic breakdown** is widespread, producing hoarding, black markets, and violence. However, struggles for survival are counterbalanced by robust pockets of local economies at a community scale. A new kind of economy emerges at the local level, focused on renewal, restoration, and regeneration.

Scenario: Imagining How the 2060s Unfold

A majority of humanity recognizes that we have reached a choice-point in history. The nurturing Earth that supported the rise of a global civilization has been transformed. Whether the biosphere can be repaired sufficiently to support the rise of a new kind of human civilization remains uncertain. The starting gun of history has gone off and we are in a race to move beyond a disaster of our own making.

Step by step, a transformed species mind emerges with a recognizable character and temperament. We progressively develop a new level of collective maturity and compassion that rises above the separations of the past. In stepping back and seeing ourselves as a contentious and yet creative species that has enormous untapped potentials for innovation and kindness, we give birth to a functioning species-civilization. An Earth-sized species-organism emerges and, with growing solidarity, we choose Earth as our enduring home. At the cost of unspeakable suffering and sorrow, we have burned through the isolations of the past to discover a deep soulful relationship with the Earth, all of her creatures, and other humans.

The underlying, creative intelligence and immense patience of the living universe becomes increasingly evident to us. We cross a threshold to new levels of collective understanding of our evolutionary journey. The entire history of our species has brought us to this opening to a larger identity, a larger humanity, and a larger future. We begin to see ourselves as cells in the body of a super-organism. As the old world unravels and falls apart, a new humanity self-assembles from these fragments.

Waves of communication envelope the Earth. A *Community Voice* movement takes root in metropolitan regions around the planet and creates a robust, grass-roots voice for humanity. "Neighborhood Voice" initiatives contribute to the bio-regional scale initiatives, lighting up the collective consciousness of the species with intense communication around the Earth. These vibrant sources of local communication merge into regional initiatives around the world. With empowered communication lighting up most regions of the Earth, a strong foundation for *Earth Voice* grows and deepens.

Divisions of race, wealth, gender, religion, ethnicity, and geography remain. However, the global communications revolution has become a powerful force for reconciliation. Martin Luther King, Jr., said that to realize justice in human affairs, "injustice must be exposed, with all of the tension its exposure creates, to the light of human conscience and the air of national opinion before it can be cured." [135] Global injustice and inequities have flourished in the darkness of inattention and ignorance. Now the healing light of public awareness creates a new consciousness among the human community. Because surviving humans know the whole world is watching, a powerful restorative and healing impulse permeates human relations. With countless resolutions, petitions, declarations, and polls from every region and level of the world, the people of the Earth make their sentiments known—we choose, again and again, to transcend our many differences and pull together in cooperation. A commitment to a regenerative and purposeful future solidifies—visibly, consciously, and deeply—in our collective psyche. Pushed by dire necessity and pulled by compelling opportunity, the great turning that humanity has been seeking gradually emerges from the grief and sorrow of the pivotal decades.

Billions of humans died in moving through the initiation of our species into early adulthood. We vow to make their sacrifice holy, never to forget; instead, we make it a sacred gift of gratitude as we learn to live within a larger aliveness. The darkness of death has ignited the flame of soulful aliveness. While still mourning the loss of so many lives, so many cultures, so many species, step by step we commit to new ways of living that honor all that has been lost—transmuting great suffering into new ways of being together.

Exhausted by the shallow projects of consumerism, we are exhilarated by the deep projects of learning to live in our living universe. We have looked squarely into the possibility of our functional extinction and, instead, have reached for a larger life. We accept our fate—recognizing there is no final truce or lasting harmony—and, instead, commit to goodwill and cooperation every day—forever.

> *Realizing there is no final rest and that we have the skills and stamina for the ongoing journey, we rise to a new level of collective awareness, maturity and responsibility.*

With a collective "yes," those who have survived make the powerful choice to find a new path forward. We commit to choosing Earth as our home for the deep future. Our long-term future is far from secure, but we are committed to the task of restoring our deeply wounded world and establishing ourselves as a viable species and civilization. A mature capacity for ethical behavior grows within us. Building on a foundation of conscious reflection and reconciliation, the human community begins the restoration and renewal of the biosphere as a common project, and this promotes a deep sense of kinship and connection. A global culture of kindness emerges.

Living in the present moment with the direct experience of being alive becomes the core source of meaning and purpose. We choose to move beyond endless pursuits of consumerism to the wealth of simply being alive in this remarkable universe. Together, we shift from a mindset of disconnection and exploitation in a dead universe to one of connection and care in a living universe.

2070s: The Great Journey– An Open Future

Summary

Looking ahead, all three of the primary pathways are still present in the world. Which of these will ultimately prevail remains unclear.

The entire Earth is still in the midst of a whole-systems crisis and the need for strong and coordinated action is so great that, without citizens stepping forward with high levels of self-organizing action, the extreme need for rapid and focused decision-making could make authoritarianism the dominant political reality.

Although the center of social gravity has shifted in favor of a transformational path forward, the threat of functional extinction of humanity remains a realistic possibility. New technologies might aid us, but will not save us. Invisible factors—such as communication, consciousness, reconciliation, aliveness—will determine the outcome.

After a half-century of turmoil and transition, we see, with unyielding clarity, that we still have three, very different, futures before us:

- Functional extinction and a new dark age.
- Authoritarian domination and evolutionary stagnation.
- Transformation and a new burst of creative evolution.

These lines from T.S. Eliot speak volumes:

> We shall not cease from exploration,
> and the end of all our exploring
> will be to arrive where we started
> and know the place for the first time.[136]

Although the path ahead remains open, the center of social gravity has shifted decisively in favor of a transformational future and the prospect of growing an ever-more mature planetary civilization. As we continue to learn, grow, and awaken, the future remains a matter of collective choice. We have not healed the great injury to the Earth. We have not established ourselves in a miraculous, new golden age of peace and prosperity. We continue struggling for survival, coping with immense challenges of global warming, the immense grief and sorrow of the great dying, the extreme difficulties of settling millions of climate refugees, restoring as many of the plant and animal species as we can, and completing the colossal challenge of transitioning to a renewable energy future. Yet what we have accomplished is momentous: We have reached a stage of mature, collective understanding as a diverse and still-contentious species. We know we must work together, forever, if we are to not perish from the Earth; now, we must find a way to live in balance with the ecology of the Earth and the living universe.

PART IV

Uplifts for a Transforming Future

It's 3:23 in the
morning
and I'm awake
because my great-great grandchildren
won't let me sleep.
My great-great grandchildren
ask me in dreams
What did you do while the planet was plundered?
What did you do when the Earth was unraveling?
Surely you did something
when the seasons started failing?
as the mammals, reptiles, birds were all dying?
Did you fill the streets with protest
when democracy was stolen?
What did you do
once
you
knew?

—*Hieroglyphic Stairway* by Drew Dellinger[137]

Uplifts for Transformation

When we heal the Earth, we heal ourselves.
—David Orr

Uplift happens when *all* of life is uplifted! To choose the *well-being of all life* as the foundation for our well-being as a species requires a profound expansion and deepening of our engagement with life. A great transition from deep separation to conscious communion, serving the well-being of all life, will not happen automatically. This is a demanding process both individually and collectively.

When confronted with the prospect of humanity's extinction, discovering forces that, if consciously chosen, can lift us up on our evolutionary journey is a treasure beyond price. Below are seven uplifting forces that are simple, universal, emotionally powerful, and can awaken our higher human potentials. Portions of these have been woven into the preceding scenario of the coming half century. Here, uplifts are explored at greater length to reveal the powerful updraft they can bring to the human journey.

1. Choosing Aliveness
2. Choosing Consciousness
3. Choosing Communication
4. Choosing Maturity
5. Choosing Reconciliation
6. Choosing Community
7. Choosing Simplicity

Let's now review each of these in more depth.

Choosing Aliveness

*The universe is a single living creature
that contains all living creatures within it.*
—Plato

*We are souls dressed up in sacred biochemical garments
and our bodies are the instruments through which
our souls play their music.*
—Albert Einstein

Soaring uplift can happen naturally when we make our home in a paradigm of aliveness that offers a new understanding of the nature of *reality* and human *identity*—and when these, in turn, bring new insights into our *evolutionary journey*. Paradigm changes that awaken this three-fold transformation are extremely rare in history. We are now in the midst of such an awakening whose essence can be summarized as *moving from deadness to aliveness*: Instead of regarding the universe as composed of dead matter and empty space without meaning or purpose, the universe is known and experienced as a unified sentient organism—a singular and living entity—becoming ever-more conscious and generating more and more complex expressions of its aliveness.

The view that we live in a unified and living universe is not "new." Instead, this is humanity's original understanding of reality, but has been largely forgotten for the past few hundred years. Now it is being rediscovered with the convergence of insights from the frontiers of science and the world's most ancient wisdom traditions.

Human's earliest intuitions revealed a subtle aliveness that permeates all existence. For at least 5,000 years, this was the view the Ohlone tribe of Indians, now extinct, but who lived sustainably on their land in the San Francisco Bay Area. Cultural anthropologist, Malcolm Margolin has beautifully described how, for the Ohlone, nature was alive and shimmering with energy. [138]Aliveness was not remote, but, like the air, was present everywhere and in everything. Because everything was filled with life, every act was spiritual. All tasks—hunting an animal, preparing food, or making a basket—were done with a feeling for the surrounding world of life and power. The perception that we live in a living universe was not restricted to indigenous cultures. More than two thousand years ago, Plato wrote his story of creation—*Timaeus*—and described the universe or cosmos as a singular, living being endowed with a soul.

Despite these deep roots of aliveness, the idea of non-living universe and dead materialism took root roughly 300 years ago in Western societies. Materialism views dead matter and empty space as the only true reality and regards the universe as being without aliveness or deeper meaning and purpose. This shallow and impoverished view of reality, human identity, and our evolutionary journey has been immensely powerful for a simple reason—it transformed the world into a resource to be consumed. If nature were essentially dead matter, then it was logical to consume the deadness to benefit the living—ourselves. This simple logic was ruthless in giving permission for the unrestrained exploitation of nature. Given the absence of ethical restraint, the paradigm of dead materialism has been merciless in its exercise of power—continuing full force until reaching the limits of its shallow and simple-minded understanding of existence. That limit is now in view, as we see the suicidal logic of dead materialism leading to the extinction our species, along with much of the rest of life on the Earth. We now confront the paradox of great impoverishment as the cost of material abundance. We are killing ourselves. The destruction of eco-systems pushes us to remember our earliest understanding of existence and to reclaim its ethical foundation: If the world around us is alive, then our mature task is to extend conscious care to all that is living and to treat it with great respect.

There is a stark and simple difference between these two paradigms: If the world is dead at the foundations, then exploit it, use it up, and consume it. If it is alive, take care of it, and use its gifts with gratitude and moderation. The modern mind has viewed nature as dead and, therefore, insentient. In turn, we have only surface regard for how we use (and abuse) it. With disregard and distance, the richness and depth of the world have been flattened into resources to exploit. Whatever uplift exists in the mechanistic paradigm amounts to a thin veneer of happiness based on consumption of more material things.

In contrast, a paradigm of aliveness abounds with uplift. Our entire universe emerged from a pinpoint of energy nearly 14 billion years ago and has blossomed into existence with an estimated two-trillion galaxies, each with a hundred-billion or more star systems! Our existence is an amazing illustration of uplift, as we continuously arise from a generative ground of aliveness. An extraordinary life-force is both *foundational* (in giving birth to and sustaining our universe) and *emergent* in giving birth to countless expressions of aliveness. We see irrepressible aliveness everywhere: for example, in grass growing through cracks of sidewalks, in the frigid reaches of the arctic ocean, in the scalding heat of deep ocean vents, in beds of clay miles beneath the Earth that have never seen sunlight and water. Sustaining an entire universe and giving birth to countless expressions of life represents astonishing uplift. Awakening to aliveness, we rediscover the continuous uplift at the foundations of all existence. If cosmic-scale aliveness can create and sustain trillions of galaxies, then it can surely provide uplift to transform the sorrow of materialism's ruin of the Earth into the joys of living in a flourishing garden, rich with possibility.

The Power of 'Aliveness'

Our collapsing world challenges us with an unyielding question: "Is there an experience of life so widely shared that it can draw us together in a common journey to a thriving future?" The answer is a straightforward "Yes." Beneath our many differences, we all share the experience of simply being alive, and this remarkable experience provides an unshakable foundation for humanity to come together in a common journey of transition and transformation.[139]

When our personal aliveness becomes transparent to the aliveness of the living universe, experiences of wonder and awe emerge naturally. As we open into the cosmic dimensions of our being, we feel more at home, less self-absorbed, more empathy for others, and an increased desire to be of service to life. These shifts in perspective are immensely valuable for building a sustainable and purposeful future.

One of the world's foremost scholars of humanity's wisdom traditions was Joseph Campbell. I had the privilege of co-authoring a book with him, *Changing Images of Man*, exploring the deep archetypes drawing us into the future in these transitional times.[140] In a revealing interview, Campbell was asked if the deepest quest of humans is the "search for meaning." He replied:

> "People say that what we're all seeking is a meaning for life. I don't think that's what we're really seeking. I think that what we're seeking is an experience of being alive, so that our life experiences on the purely physical plane will have resonances with our own innermost being and reality, so that we actually feel the rapture of being alive."[141]

A quote attributed to philosopher Blaise Pascal speaks clearly: "The goal of life is not happiness, peace, or fulfillment, but *aliveness*."[142]

Howard Thurman, author, philosopher, theologian, and civil rights leader famously said, "Don't ask what the world needs. Ask what makes you come alive and go do it. Because what the world needs is people who have come alive."[143]

Aliveness is our only true wealth

Psychologist and philosopher, Erich Fromm, has written that our experience of aliveness is the most precious gift we can share with others. When we share the experience of aliveness within ourselves—our gratitude and fears, understanding and curiosity, humor and sorrow—we offer the essence of our being. In sharing our aliveness, we enrich the life of others. We awaken their sense of aliveness by sharing our own experience of being alive in the moment. We do not share with the intention of receiving something from others; instead, the sharing itself is a gift of ourselves that awakens a reciprocal aliveness in others, returning us in a mutually enhancing flow.

Spiritual and ecological elder Joanna Macy connects climate activism with our experience of aliveness:

> "The current moment is an exquisite time to be alive. Because, an awareness of impending collapse is an invitation to ask ourselves deep questions of meaning that we typically postpone—and some of us never even get to. *Climate despair is inviting people back to life.* . . . The way through despair involves experiencing oneself as part of a greater whole and surrendering to the mystery of creation. . . . The climate crisis invites us to engage with the mystery of life with fresh eyes and an open heart."[144]

Jungian philosopher Anne Baring describes how consumer cultures find it difficult to enter into the experience of indigenous cultures and their understanding that: ". . . the life of the Cosmos, the life of the Earth, and the life of humanity were one life, permeated and informed by animating spirit."[145] She writes that the great revelation of our time is that "we are moving from the story of a dead, insentient cosmos to a new story of a Cosmos that is vibrantly alive and the primary ground of our own consciousness."[146]

A non-living universe is without consciousness and is therefore oblivious to any sense of human purpose. As existentially separate life-forms, we may strive heroically to impose some reason for our existence on the universe, but this is ultimately fruitless in a cosmos unaware of life. In striking contrast, a living universe seems intent on generating self-referencing and self-organizing systems within itself at every scale. We are expressions of aliveness that, after nearly 14 billion years, enable the universe to look back and reflect upon itself. A living universe paradigm brings a profound shift in our evolutionary purpose:

> *Life is occupied in both perpetuating itself and surpassing itself; if all it does is maintain itself, then living is only not dying."*[147]
> **—Simone de Beauvoir**

Beneath differences of language and history, a common understanding is present—the universe is a living system emerging as a fresh creation in every moment. We are an inseparable part of this regenerative process. This understanding is well-known and recognized widely by mystics, poets, and naturalists:[148]

> *Heaven is under our feet as well as over our heads.*
> **—Henry David Thoreau**[149]

> *The deeper we look into nature, the more we recognize that it is full of life. . . . From this knowledge comes our spiritual relationship with the universe.*
> **—Albert Schweitzer**[150]

> *And into the forest I go to lose my mind and find my soul.*
> **—John Muir**[151]

> *Not just beautiful, though—the stars are like the trees in the forest, alive and breathing. And they're watching me.*
> **—Haruki Murakami**[152]

> *The goal of life is to make your heartbeat match the beat of the universe, to match your nature with Nature.*
> **—Joseph Campbell**[153]

> *If you wish to know the divine, feel the wind on your face and the warm sun on your hand.*
> **—Buddha**[154]

> *I believe in God, only I spell it Nature.*
> **—Frank Lloyd Wright**[155]

Awakening to our conscious connection with the living universe naturally expands our scope of concern and compassion—and brightens the prospect of working together to build a sustainable future. This is not abstract philosophy, but the visceral experience of simply being alive to our unique experience of ourselves. At 90 years of age, the words of Florida Scott-Maxwell describe this view powerfully: "You need only claim the events of your life to make yourself yours. When you possess all you have been and done, you are fierce with reality."[156]

As we awaken to the aliveness at the core of our being, we simultaneously connect with the aliveness of the universe.

Aliveness costs nothing and is freely given as our birthright. The experience of aliveness is here and available to us at all times. Aliveness is an embodied, powerful, and universally shared experience. To illustrate, I asked participants in a learning community I co-facilitate to describe what "being fully alive" means to them. Responses were immediate and direct: "Being in the flow." "Mind coming home to body." "Feeling the full range of my emotions." "Living on purpose and without expectation." "Giving full expression of my soulful gifts." "Deep connection with nature."[157]

A life path devoted to developing full aliveness may be dismissed as wishful fantasy by those living within the mindset of materialism and consumerism. However, this view is changing. The mindset of materialism is being transformed by new findings from science, by enduring insights from wisdom traditions, and by the direct experience of a large portion of humanity. By integrating these diverse sources of understanding, we discover aliveness is the new—and ageless—experience that offers humanity a place of common meeting and collective healing.

Our closest connection with the earliest understandings of ancient peoples comes from indigenous traditions with deep roots extending far into humanity's past. Indigenous wisdom supported our ancestors as they endured exceptionally harsh conditions for several hundred- thousand years. How do people who continue to uphold these ancient traditions experience life and the world?

The Koyukon Tribe of northern central Alaska
The Koyukon live "in a world that watches, in a forest of eyes." They believe wherever we are, we are never truly alone because the surroundings, no matter how remote, are aware of our presence and must be treated with respect.[158]

Sarayaku Kichwa, of the Ecuadorean Amazon jungle
Believe that "Everything in the jungle is alive and has a spirit."

Luther Standing Bear, Lakota Sioux from the region of North and South Dakota
"There was no such thing as emptiness in the world. Even in the sky there were no vacant places. Everywhere there was life, visible and invisible, and every object gave us great interest in life. The world teemed with life and wisdom; there was no complete solitude for the Lakota."[159]

The idea and experience of a living conscious presence permeating the world is shared by most (perhaps all) indigenous cultures. The Alaskan Koyukon people described the natural world as a "forest of eyes" aware of our presence, no matter who or where we are. A related intuition tells us that a life-force or "sacred wind" blows through the universe and brings with it a capacity for awareness and communion with all life.

Consistent with indigenous views, we find an astonishing insight regarding the nature of the universe in diverse spiritual traditions. Most spiritual traditions view the universe as continuously arising anew at every moment—an undivided whole emerging in an unutterably vast process of awesome precision and power:

Christianity: *"God is creating the entire universe, fully and totally, in this present now. Everything God created ... God creates now all at once."*[160]
—Meister Eckhart, Christian mystic

Islam (Sufi): *"You have a death and a return in every moment.... Every moment the world is renewed, but we, in seeing its continuity of appearance, are unaware of its being renewed."*[161]
—Jalāl ad-Dīn Muhammad Rūmī, 13th century Sufi teacher and poet

Buddhism (Zen): *"My solemn proclamation is that a new universe is created every moment."*[162]
—**D. T. Suzuki, Zen teacher and scholar**

Hinduism: *"The entire universe contributes incessantly to your existence. Hence the entire universe is your body."*[163]
—**Sri Nisargadatta, Hindu teacher**

Taoism: *"The Tao is the sustaining Life-force and the mother of all things; from it, all things rise and fall without cease."*[164]
—**Lao Tzu, founder of Taoism**

How widespread is the experience of permeating aliveness and deep unity in everyday life? How often do people feel aliveness and intimate connection with nature and the larger world? Scientific surveys have explored this pivotal question:

- A global survey involving 7,000 youths in 17 countries taken in 2008 found that 75 percent believe in a "higher power," and a majority say they have had a transcendent experience, believe in life after death, and think it is "probably true" that all living things are connected.[165]

- In 1962, a Gallup survey of the adult population in the U.S. found that 22 percent reported having awakening experiences that reveal our intimate connection with the universe. By 1976, Gallup reported this had grown to 31 percent. By 1994, a *Newsweek* survey found this had grown to 33 percent. By 2009, a Pew Research survey reported that "moment[s] of sudden religious insight or awakening" had grown dramatically to 49 percent of the adult population.[166]

- In a national survey of the U.S. in 2014, nearly 60 percent of adults reported they regularly feel a deep sense of "spiritual peace and wellbeing," and 46 percent say they experience a deep sense of "wonder about the universe" at least once a week.[167]

- An important reason for these changes may be the dramatic increase in meditation in recent years. A New Age novelty in the 1960s has grown into a mainstream movement in the 21st century. The percentage of adults who meditate is growing rapidly: from an estimated four percent of the U.S. population

Figure 5: Growth in Awakening Experiences in US 1962-2009 by Percentage of Population

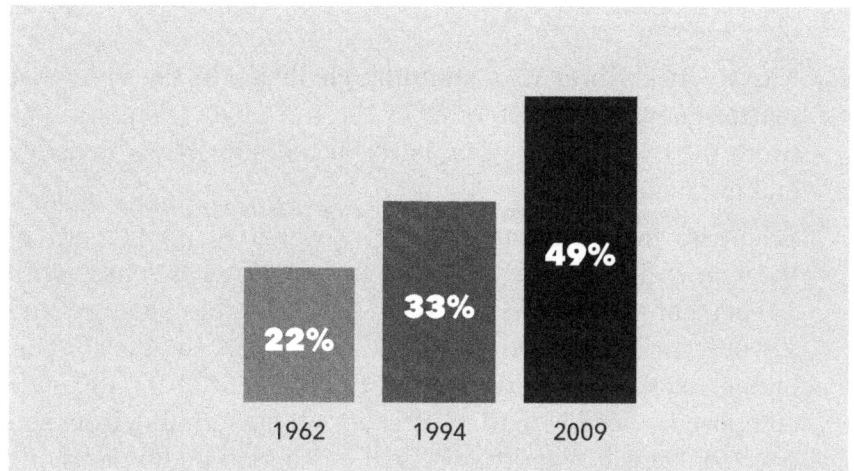

in 2012 to more than 14 percent just five years later (2017).[168] Meditation, diet, and exercise, are now considered mainstream activities for health and well-being.

These surveys show that awakening experiences of communion and connection with the aliveness of the universe are not a fringe phenomenon, but are familiar for a large portion of the public. Humanity is measurably waking up to a view of ourselves as inseparable from the larger universe.[169]

Until recent decades, any suggestion that the universe could be viewed as a unified living system was regarded as fantasy by mainstream science. Now, with findings from quantum physics and elsewhere, the ancient intuition of a unified living universe is being reconsidered, as science cuts away superstition to reveal the cosmos as a place of unexpected wonder, depth, dynamism, and unity.[170]

- **A Unified Whole**: In the last several decades, quantum physics has repeatedly confirmed that the universe is single, vast unity deeply connected with itself everywhere at every moment. A famous quote from Albert Einstein challenges the view of separation: "A human being is part of the whole called by us 'universe,' a part limited in time and space. We experience ourselves, our thoughts, and feelings as something separate from the rest. A kind of optical delusion of consciousness. The quest for liberation from this bondage is the only object of true religion."[171]

- **Mostly Invisible**: In a stunning challenge to the view that matter-energy is all there is in the universe, scientists now think the overwhelming majority of the universe is invisible and not material!

Scientists now estimate that approximately 95 percent of the known universe is invisible to our physical senses with 72 percent comprised of "dark" (or invisible) energy and 23 percent comprised of "dark" (invisible) matter.[172] Our biology is a manifestation of the four percent of the universe composed of visible matter. This new understanding from science confirms humanity's original perception that, underlying the physical world, there is a vastly larger invisible world of unseen energy and immense power.

Here is an even more far-reaching view from Albert Einstein: "What we have called matter is energy, whose vibration has been so lowered as to be perceptible to the senses. Matter is spirit reduced to point of visibility. There is no matter."

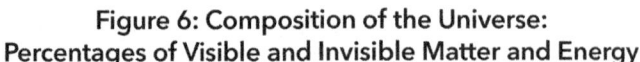

Figure 6: Composition of the Universe:
Percentages of Visible and Invisible Matter and Energy

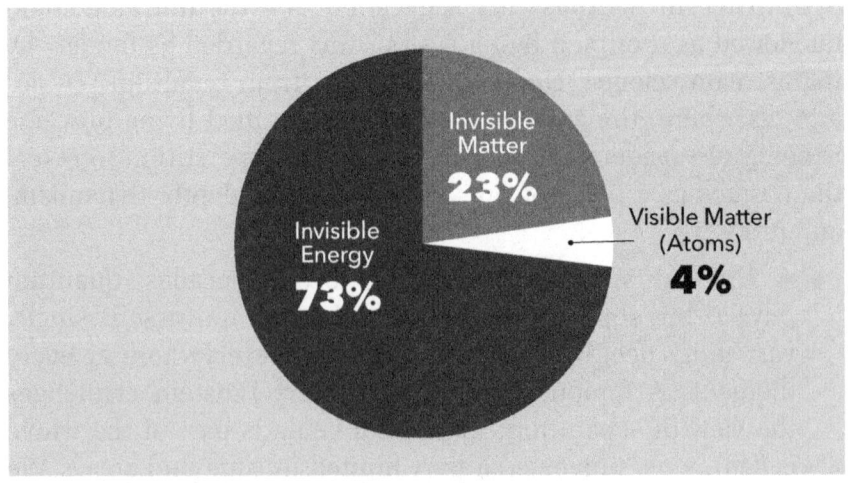

- **Co-Arising**: At every moment, the entire universe emerges freshly as a singular orchestration of cosmic expression. Nothing endures. All is flow. In the words of cosmologist Brian Swimme, "The universe emerges out of an all-nourishing abyss not only fourteen-billion years ago, but in every moment."[173] Despite outward appearances of solidity and stability, when science explores deeply, we see evidence that the universe is a regenerative system.

- **Consciousness at Every Scale**: A gradation of consciousness seems to be present throughout the universe so that consciousness never switches off completely as we explore ever smaller expressions of life; instead, consciousness diminishes as organic complexity reduces—from humans to dogs, insects, plants, and single-celled creatures, and then continues fading into inorganic matter such as electrons and quarks that possess an extremely simple form of consciousness congruent with their simple nature.[174] Further, because the universe is a unified whole and there are no independent parts, it suggests the universe itself has consciousness, an expression of its holistic nature, and that can be experienced by humans as the consciousness of the cosmos or "cosmic consciousness."[175] The generative power of the "Mother Universe"—which has given birth to our "daughter universe"—suggests there exists an underlying ocean of generative aliveness and awareness from which an entire universe can spring, and grow from a seed smaller than a single atom to a vast system with several trillion galaxies. Max Planck, developer of quantum theory, stated, "I regard consciousness as fundamental."[176]

- **Able to Reproduce**: A vital capacity for any living system is the ability to reproduce. A growing view in cosmology is that our universe reproduces itself through black holes. Physicist John Gribbin writes, "Instead of a black hole representing a one-way journey to nowhere, many researchers now believe that it is a one-way journey to somewhere—to a new expanding universe in its own set of dimensions."[177]

A new picture of our universe is coming into focus. Life exists within life. Our aliveness is inseparable from the larger aliveness of a living cosmos. The universe is a unified "super-organism" continuously regenerated at each moment, and that includes consciousness, a knowing capacity, that enables systems at every scale of existence to exercise some measure of freedom of choice.

We are not who we thought we were. In considering the enormity of the universe with its billions of swirling galaxies, each with billions of stars, it is natural to draw the conclusion that we are utterly tiny in the cosmic scale of things. However, this view is radically mistaken. We are not small creatures—in the overall scale of the universe, we are literally giants! Imagine that you have a ruler that measures from the largest scale of the known universe to the smallest. At the largest scale, we see hundreds of billions of galaxies and, at the smallest scale, we travel deep within the core of an atom and then much further down to the unimaginably small realms of our regenerative universe. If we place the size of humans alongside this cosmic ruler, we find that we are in the middle range. In fact, *there is more smallness within us than bigness beyond us*! In the cosmic scale of things, we are truly enormous creatures—we are giants! As colossal beings, it is easy for us to overlook the whirlwinds of regenerative activity continuously at work at the truly microscopic scale of the universe.

Thomas Berry, scholar of the world's religions, describes the inseparable connection of the individual with the universe: "We bear the universe in our being as the universe bears us in its being. The two have a total presence to each other and to that deeper mystery out of which both the universe and ourselves have emerged."[178] How extraordinary: A field of aliveness creates and sustains our universe, patiently holding it within its spacious embrace for billions of years, while producing ever-more conscious expressions of aliveness increasingly able to look back with reflective consciousness and appreciate their origins.

As we learn to recognize our experience of aliveness, and as we encounter the aliveness at the foundation of the universe as felt experience—as life meets life—a window opens and awakening experiences naturally arise. When our experience of aliveness connects with the greater aliveness of the universe, we recognize, as direct experience, that we are part of the great wholeness of life. This is who we are: both a unique biological aliveness and an inseparable part of cosmic aliveness. We are both biological and cosmic in nature—we are "bio-cosmic" beings. In a stunning paradox, as we grow in our spiritual maturity and become one with all life, we simultaneously become ever-more completely and uniquely ourselves.

When we bring these many threads of wisdom together—indigenous understandings, spiritual traditions, nature wisdom, direct experience, and scientific evidence—they transform our understanding of *reality* (from dead to alive), and this transforms our understanding of human *identity* (as both biological and cosmic in nature), and this transforms our understanding of our evolutionary *journey* (we are learning to live in a living universe).

To review: The paradigm of materialism assumes we inhabit a universe that is non-living at its foundations, without consciousness, meaning, or purpose. As a result, we identify with our material or biological nature, and no more. I think and, therefore, I am the thoughts that I think and no more. In contrast, in a living universe, our being includes consciousness that reaches into an unbounded ecology beyond our thinking brain. Therefore, as conscious beings, our identity can reach far beyond our biological nature and mental activity. We are beings of both biological and cosmic dimension—to repeat: *we are bio-cosmic beings*. Just as we can cultivate and evolve our thinking capacity, we can also evolve our capacity for unbounded knowing in the unity of the universe. The expansion and deepening of our natural capacity for cosmic consciousness transforms our identity and evolutionary journey.

Yet being unflinchingly realistic, it does not seem likely we will turn away from a path of separation—with its growing inequities, overconsumption of resources, and deep injury to the Earth—unless we discover a pathway to the future so truly remarkable, transformative, and welcoming that we are drawn together by the felt presence of its invitation. This path must be so compelling as a felt possibility that we are drawn into exploration in the present moment. That pathway is being revealed by insights converging from science and the world's wisdom traditions.

> *We are discovering that, instead of struggling for meaning and a miracle of survival in a dead universe, we are invited to learn and grow forever in the deep ecologies of a living universe.*

Stepping into the invitation of learning to live in a living universe represents a journey so extraordinary, it calls us to heal the wounds of history and to realize a remarkable future we can attain only together. As we open into the cosmic dimensions of our being, we feel more at home, less self-absorbed, more empathetic toward others, and increasingly drawn to be of service to life. These shifts in perspective are immensely valuable for building a sustainable future.

> *To step into the invitation of learning to live consciously in a living universe is to begin a new chapter in humanity's evolution with a transformed understanding of reality, human identity, and our evolutionary journey.*

If only for brief moments at a time, we *can* glimpse and know existence as a seamless totality. Touching the aliveness of the universe for even a few moments can transform our lives. The deeply loved Sufi poet Kabir wrote that he saw the universe as a living and growing body "for fifteen seconds, and it made him a servant for life."[179] No matter how mundane the circumstance, no matter how seemingly trivial the situation, we can always become aware of the subtle aliveness and consciousness within and around us. We can glimpse the living universe in the golden light of a late afternoon or in the luster of an old wooden table that shines with an inexplicable depth and presence. We can also witness the buzzing aliveness of existence in places that may seem far removed from nature—a room filled only with plastic, chrome steel, and glass will fiercely display aliveness in the raw. In the gentle contemplation of any part of ordinary reality, we can catch glimpses of the great hurricane of energy that blows with silent force through all things and, with a "forest of eyes," is aware of our existence. Empty space will also disclose that it is an ocean of dancing aliveness—a subtle symphony of transparent architecture actively providing a context for matter to present itself.

To be born as a human being is a rare and precious gift. While we have the gift of a body to anchor our experience, it is important to recognize our bio-cosmic nature.

> *We are bio-cosmic beings:*
> *Our bodies are biodegradable vehicles*
> *for acquiring soul-growing experiences.*

As compostable conduits for cosmic learning experiences, our bodies are expressions of a creative aliveness that, after nearly 14-billion years, enables the universe to look back and reflect on itself. Because the cosmos is a learning system, a primary purpose for being here is to learn from both the pleasures and the pains of existence. If there were no freedom to make mistakes, there would be no pain. If there were no freedom for authentic discovery, there would be no ecstasy. In freedom, we experience both pleasure and pain in the process of developing our identity as beings of both earthly and cosmic dimensions.

We stand upon the Earth as agents of self-reflective and creative action engaged in a time of great transition, consciously learning to live in a living universe. An ancient Greek saying speaks directly to our learning journey: "Light your candle before night overtakes you." If the universe were non-living at its foundation, it would take a miracle to save us from extinction at the time of death, and then to take us from here to a heaven (or promised land) of continuing aliveness. However, if the universe is alive, then we are already nested and growing within its aliveness.

> *All things end.*
> *All being continues.*
> *That is the nature of each.*

When our physical body dies, the life-stream we are makes its passage to a fitting home in the larger ecology of aliveness. We don't need a miracle to save us—we already exist within the miracle of sustaining aliveness. Instead of being saved from death, our job is to bring mindful attention to the ever-emerging aliveness in the here and now. We are shifting from seeing ourselves as accidental creations wandering through a lifeless cosmos without meaning or purpose to seeing ourselves engaged in a sacred journey of discovery in a living cosmos of stunning depth and richness of purpose. Cynthia Bourgeault, a modern-day mystic and Episcopal priest, writes, "Each one of us, and every action we make, has a quality of aliveness to it, a fragrance or vibrancy uniquely its own. If the outer form of who we are in this life is conveyed by our physical bodies, the inner form—our real beauty and authenticity—is conveyed in the quality of our aliveness. This is where the secret of our being lies."[180]

In learning to live in a living universe,
we are learning to live in the deep ecology of existence.
This is such an astonishing call to our soulful nature
from the deep compassion of a living universe
that we would be cosmic fools
to ignore an invitation whose value
is beyond price or measure.

An old saying goes, "a dead man tells no stories." In a similar way, "a dead universe tells no stories." In contrast, a living universe is itself a vast story continuously unfolding with countless characters playing out gripping dramas of awakening and creative expression, inseparable from the artistry of world-making. The universe is a living, unfolding creation. Saint Teresa of Avila saw this when she wrote, "The feeling remains that God is on the journey, too."[181] If we consciously recognize ourselves as participants in a cosmic garden of life that has been growing patiently over billions of years, we can awaken to the soaring uplift of aliveness and move from feelings of cosmic separation to feelings of cosmic participation, curiosity, and love.

Choosing Consciousness

"*In the history of the collective, as in the history of the individual, everything depends on the development of consciousness.*"
—**Carl Jung**

Ancient wisdom suggests there are three miracles in life. First, that anything exists at all. Second, living things (plants and animals) exist. Third, living things know they exist. The third miracle is capacity for self-reflective consciousness and is fundamental to our nature as humans. Our scientific name is *Homo sapiens sapiens*—we are not only "sapient" (beings with the ability to know), we are beings who can "know that we know" and can watch or witness ourselves as we move through our daily lives. We see that when not running on automatic—not following habitual and pre-programmed ways of living—we have freedom of choice. Consciousness and freedom are intimate partners in the dance of evolution. Reflective consciousness is a powerful aid for uplift and movement through this time of initiation for our species.

The first step in uplift and evolution is to simply see "what is"—to become an impartial observer or witness of our own experience. Honest reflection and nonjudgmental witnessing are fundamental to uplifting our lives. Through paying attention to our lives in the mirror of consciousness, we can make friends with ourselves, and come to greater self-possession. The capacity for honest self-reflection helps to cut through the surface chatter of our lives and discover the direct experience of our existence.

Peter Dziuban writes about the relationship of consciousness and aliveness.[182] He describes "aliveness" as a direct experience, instead of something we think about. He asks us to imagine a wine-tasting party where the tasting is the purpose. So it is with life. We are here to taste what it means to be alive—to directly experience and live our aliveness. Dziuban writes, "Life is nothing if it is not alive!" In the simplicity of silence, we can taste aliveness. Our aliveness is not a thought, but a living presence. Nor is aliveness a thought *about* aliveness—it is the *direct experience of aliveness itself.*

> "You are conscious and alive. The words and thoughts are what you are *conscious of*. Words and thoughts by themselves are never conscious—only you are. So that's what you really are, this pure consciousness—not unconscious words and thoughts *about* it. Huge difference. Thinking is a changing process. Aliveness is a changeless Presence."[183]

To witness or watch ourselves moving through life is not a mechanical process, but a living experience where we consciously "taste" our lives, and make friends with ourselves, including those moments of doubt, anger, fear, and desire we may prefer to ignore. An "observer self" or "witnessing self" gives us the ability to stand back from complete identification with bodily-based desires, emotions, and thoughts. With the trustworthy mirror of reflective consciousness, we can see ourselves as if from a distance. From this perspective, we see that, while our bodily experience is one part of ourselves, we are more than our body's sensations, pleasures, and pains. We also see that while emotional experience is one part of us, we are more than our experience of anger, happiness, and sorrow.

By bringing reflective consciousness into our lives, we experience more spaciousness and freedom. We no longer identify exclusively with sensations, emotions, and our inner stream of mental dialogue. The detachment and perspective provided by reflective knowing supports the reconciliation needed for moving through this time of great transition. When present with reflective consciousness, we no longer operate largely on automatic. Expanding reflexive awareness to a social scale—seeing ourselves in the mirror of mass media (the internet, television, and other tools of the global nervous system)—changes everything. Recognizing that we live in a shared ecology of consciousness weaves the human family into a mutually appreciative whole, while simultaneously honoring our differences.

Reflective consciousness is vital for coping with intense global stresses and challenges. We have entered a perfect storm of intertwined and critical problems that require an unprecedented level of global reflection and reconciliation inspired by a shared vision of a sustainable future. Here is how eminent scientist Carl Sagan expressed our situation when testifying before Congress in 1985 on how the greenhouse effect will change the global climate system:

> "What is essential for this problem is a global consciousness. A view that transcends our exclusive identifications with the generational and political groupings into which, by accident, we have been born. The solution to these problems requires a

perspective that embraces the planet and the future because we are all in this greenhouse together."[184]

Importantly, the awakening of consciousness does not end with mindfulness or reflective attention. Beyond reflective consciousness and the polarity of watcher and watched, or observer and observed, we can evolve to unitive consciousness. If we persevere with sustained mindfulness, the distance between observer and observed gradually diminishes until we become a single, integrated flow of experience. As the knower and that which is known converge and become one in experience, we realize we are inseparable from whatever we observe. Because the universe is a profoundly unified whole, we simply allow our conscious knowing to coincide with what is known. We let go of objectifying reality as something to be witnessed "out there," and realize that reality can be directly experienced "in here." We can move beyond "reflecting on" life and move into the experience of "coinciding with" (or simply *being)* life.[185]

A new social atmosphere will emerge in a culture of compassionate consciousness. No matter where people are in the world, we will increasingly know we are among relatives. Our sense of identity will expand and we'll regard everyone as "compassionate citizens of the cosmos"—beings immersed in the depths of a living universe, who feel a deep kinship with all life.

The word "passion" means "to suffer" and the word "com-passion" literally means "to suffer with." If we watch people move through a painful transition, we can become one with the experience of suffering and will naturally work to alleviate that suffering. Swimming in the larger ocean of life, we know intuitively that if the Earth is suffering, we all bathe in an ocean of subtle suffering. We recognize that our experience of life is permeable and that we share in whatever measure of happiness or sorrow is created for the whole.

As the push of outer necessity meets the pull of untapped inner capacity, humanity awakens its capacity for conscious reflection and knowing. We recognize that if we are distracted and in denial, and overlook the urgency and importance of the great transition now underway, we will miss a unique, never-to-be-repeated, evolutionary opportunity.

Each generation makes sacrifices for the next as a caretaker for the future. The current generation is being pushed by a wounded Earth and pulled by a welcoming universe to make an unprecedented gift to humanity's future: awakening together with equanimity and maturity to consciously realize our bio-cosmic potential and purpose of learning to live in a living universe.

Witnessing or reflective consciousness moves from the status of a spiritual luxury for the few in an earlier fragmented world to that of a social necessity for the many in our modern interdependent world. The quality of our personal and social attention is the most precious resource and gift that we can offer to life. Old wisdom takes on new meaning: "The price of freedom is eternal vigilance." Our level of social vigilance is fundamental to the functioning of a free society. If we don't pay attention while decisions of evolutionary importance are being made, then we effectively forfeit our future. Now is the time to be wide awake, both personally and collectively.

To be free from needless government intrusion, individuals and communities must develop their capacity to be consciously self-regulating at a pace at least equal to that at which the social order grows more complex, interdependent, and vulnerable. Bringing reflective consciousness into our wired world enables us to objectively witness, for example, the deep wounds of racism, poverty, intolerance, gender discrimination. Observing consciousness enables us to stand back and experience our common humanity from a dispassionate perspective, providing invisible glue to bond the human family into a workable community.

Developing a more conscious, reflective society allows many other enabling capacities to emerge, including:

- **Self-Determining**—One of the most basic expressions of a maturing consciousness is an enhanced capacity for self-determination. A conscious society is able to look at its options, as well as observe itself in the process of choosing. We are able to observe our collective self "from the outside," much as one culture or nation can view another. A reflective society does not blindly trust a particular ideology, leader, or political party. Instead, it regularly reorients itself by looking beneath slogans and vague goals to choose a preferred path to the future.

- **Error-embracing**—A more conscious society recognizes that social learning inevitably involves making mistakes. Therefore, errors are not automatically regarded as "bad," instead, they are accepted as important feedback in the process of learning.
- **Equanimity**—A more conscious society tends to be objective and react calmly to the stressful pushes and pulls of trends and events. It demonstrates an evenness, levelheadedness, and confidence unmoved from its center by the passions of the moment.
- **Inclusive**—A more conscious society continually searches for synergy as different ethnic groups, geographic regions, and ideological perspectives are actively invited into a search for higher common ground.
- **Anticipatory**—In viewing the world more objectively from a larger perspective, a reflective society tends to consciously consider alternative pathways into the future. Instead of waiting passively for crises to force action, we pay greater attention and respond to danger signals.
- **Creative**—A conscious society is not locked into habitual patterns of thinking and behaving. Rather than respond with preprogrammed solutions, it explores options with a fresh and flexible frame of mind.

These qualities of an awakening reflective consciousness bring powerful uplift for moving through our time of collective initiation.

Choosing Communication

Communication is the lifeblood of civilization. The ability to communicate enabled humans to progress from gatherers and hunters to the edge of a planetary eco-civilization. Empowered by the internet and television, the human family is moving from a history of separation to a future of instantaneous global communication and connection. Each day, more than half of humanity reaches into the expanded reality of television and the internet. With stunning speed, we are developing skills for local-to-global communication that are transforming our collective communication *and consciousness* as a species. As the internet becomes faster, smarter, and more encompassing, it weaves humanity together into a single web of communication that functions like a "brain" for planet Earth.

No longer isolated from one another, we are collectively witnessing our world in profound transition. Awakenings and innovations happening on one side of the planet are communicated instantly around the world, enabling us to wake up together. With astonishing speed, humanity is rousing from its collective slumber to discover ourselves as a single species, united by an extraordinary network of planetary communication. The Earth is beginning to establish a voice for itself that transcends local and national interests.

These tools can offer humanity a clear window for seeing the world and an unwavering mirror in which to see ourselves. With the internet and television, we have tremendously powerful technologies for lifting ourselves out of denial and distraction and into a future of profound transformation. However, with authoritarian controls, these same tools can contract our social attention into a cramped and censored reality. It is important to be mindful of both possibilities. We can either rise to higher human potentials with these powerful tools of communication or descend into a dark well of digital authoritarianism.

Historically, when an authoritarian government comes to power, one of the first actions it takes is to seal off a country to prevent the free flow of communication with the outside world. Next, they clamp down on free speech and dissent within the country. Digital dictatorships that limit communication both inside and outside a country are growing throughout the world. Countries such as China and Russia are shutting down internet sites, subduing opposition, and imposing draconian prison terms for online dissent.

In other countries, such as the U.S., restrictions on media freedoms are imposed, not by the government, but by the self-censorship of media companies seeking to maximize their profits by churning out entertainment programs filled with commercial advertising. In the U.S., we can see the results of this consumer bias in the grossly inadequate levels of attention given to climate catastrophe, species extinction, and other areas of the deepening Earth crisis. To illustrate: If we add up the number of minutes of climate

Figure 7: Airtime on Broadcast TV about Climate Change: 2017, 2018, 2020

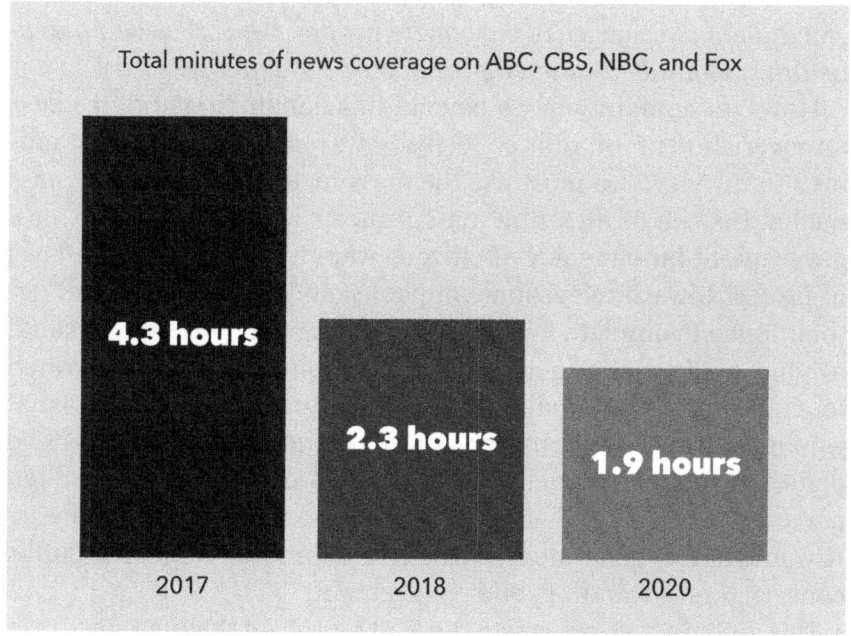

coverage from broadcast TV networks (ABC, CBS, NBC, and Fox), *for an entire year,* we see that the total number of minutes of news programming dropped from just over four hours in 2017 to just over two hours in 2018.[186] *Two hours of collective attention to our global climate crisis for an entire year! This is an astonishingly inadequate level of attention for a modern democracy facing a planetary crisis!* Other factors such as the mass extinction of species are essentially being ignored altogether.

In 2020, the overall coverage of climate-change on broadcast TV news shows plummeted even further—by 53 percent. Over the course of an entire year, these news shows covered climate change for a total of 112 minutes (less than two hours) in 2020—the lowest amount of coverage since 2016.[187] This drastic decrease in climate coverage occurred despite numerous climate-fueled extreme weather events, important reports regarding the effects of climate change, repeated assaults on the environment by political and business interests, and a presidential election in which climate change took center stage. Overall, climate coverage in 2020 made up only four-tenths of one percent of broadcast television news shows. *This abysmal level of attention illustrates with startling clarity how, in service of corporate profits, the U.S. is being devastatingly dumbed-down by broadcast TV networks.*

How can humanity move beyond this debilitating and needless impoverishment of our collective awareness and understanding? In my view, we must use the mass media to change the mass media. Instead of directing mass protests at an oil company or a government bureaucracy, if citizens were to direct the same level of protest toward television companies and stations, and call out their almost complete failure to serve the public interest, it could produce a dramatic increase in the amount of airtime devoted to exploring critical challenges to our future. For example, what would happen to public understanding of the Earth crisis if, instead of half of one percent of broadcast time, broadcasters devoted ten percent or even 20 percent of prime time to this existential threat? It would surely generate a swift and revolutionary uplift in public concern, understanding, and engagement!

It is vital that we recognize the leading role of the mass media in promoting the collective madness of materialism. It is literally insane for us to overconsume the Earth and force a descent into either digital authoritarianism or functional extinction as a species. The U.S. stands as the prime example of this madness—where the average person watches more than four hours of television each day—which means that, *as a civilization, Americans watch more than one billion person-hours of television each day.* In turn, it is estimated that the average American will watch more than 25,000 commercials a year! Commercials are much more than advertisements for products; they are highly sophisticated messages and stories that prioritize and promote materialistic values and ways of living.

There may be no more dangerous a challenge to our future than the cultural hypnosis created by commercial television, which trivializes human life and distracts humanity from our rite of passage into early adulthood. *By programming television for commercial success, the mindset of civilizations is programmed for evolutionary stagnation and ecological failure.* Media companies tell us we should consume more, while our ecological concerns for the Earth tell us we must consume less. Carl Jung said schizophrenia is a condition where "the dream becomes the reality." The American dream of consumerist lifestyles has become our primary reality—increasingly out of touch with the reality of the Earth and our evolutionary potentials. Decades ago, Professor Gene Youngblood warned of the possibility that the mass media could lock in a materialistic mindset and hold back human evolution, simply by controlling the perception of alternatives.

> "The industrial order endures not by conspiracy but simply by default, simply because there's no popular demand for a specifically-defined alternative. . . . Desire is learned. Desire is cultivated. It's a habit formed through continuous repetition. . . . But we cannot cultivate that which isn't available. We don't order a dish that isn't on the menu. We don't vote for a candidate who isn't on the ballot. . . . We rarely select what's scarcely available, seldom emphasized, infrequently presented. . . . What could be a more radical example of totalitarianism than the power of the mass media to synthesize the only politically relevant reality, specifying for most people most of the time what's real and what's not, what's important and what's not. . . ? This, I submit, is the very essence of totalitarianism: the control of desire through the control of perception. . . . What prevents our frustration from shaping new institutions is the inability to perceive alternatives, resulting in the absence of desire, hence of demand, for those alternatives."[188]

Our situation is unprecedented in history. We humans confront the trailblazing challenge of coming together on behalf of a sustainable and meaningful future for us all. Martin Luther King, Jr. described the challenge this way:

> We are challenged to rise above the narrow confines of our individualistic concerns to the broader concerns of all humanity. . . . Through our scientific genius we have made of the world a neighborhood; now through our moral and spiritual genius we must make of it a brotherhood.[189]

These are pivotal years for the communications future of humanity. Will communication—the lifeblood of our species—be weak, lackluster, and pale, or strong, creative, and colorful? How well we communicate will make a tremendous difference in whether we can mobilize sufficient uplift to rise above the downdraft of either extinction or authoritarianism.

It is helpful to recognize both the strengths and weaknesses of the two technologies at the heart of the communications revolution: television and the internet.

- Television has great breadth of reach, but is generally shallow.
- The internet has great depth of reach, but is generally narrow.

Isolated from one another, these tools generate communication that tends to be *shallow and narrow*. However, if we combine the power of each, we can awaken communication that is *deep and wide*! These are not competing technologies, but complementary and highly synergistic. The tools for a revolution in communication surround us if we will make conscious use of them.

Turning to local empowerment, we can build on more than a century of experience in the U.S. with the "New England Town Meetings" where residents of a town voted on issues of common concern. In the modern era, we can consider an entire metropolitan area (San Francisco, Philadelphia, Paris, etc.) as a "town" and the residents of that area eligible to "vote" and offer their advisory views on key concerns, such as the climate crisis.

A metropolitan-scale, electronic town meeting is not a fantasy—the feasibility of this approach was demonstrated decades ago (in 1987) in the San Francisco Bay Area. I was the co-director of a non-profit and non-partisan organization called "Bay Voice"—the electronic voice of the Bay Area. In collaboration with the ABC-TV station, we produced an hour-long, prime-time, non-partisan Electronic Town Meeting. *We understood that, in U.S., broadcast TV stations (ABC, CBS, NBC, and Fox) that use the public airwaves have a strict legal obligation "to serve the public interest, convenience, and necessity" of the community they serve, before they serve their own profit-making interests.*[190] To build our *Community Voice* organization, we brought together a diverse coalition of citizen groups—including different ethnic groups, business and labor organizations, and environmental organizations. This broad coalition genuinely represented the diverse views and interests of the Bay Area community. To produce the pilot electronic town meeting, we worked with two major universities (Stanford and UC Berkeley) we developed a scientific or random sample of citizens who could participate by giving feedback from their homes. Those who agreed were sent a list of phone numbers that corresponded to various options they could dial in (this experiment was conducted more than a decade before the internet became widely used).

The pilot "Electronic Town Meeting" (abbreviated below as "ETM") began with an informative, mini documentary to place our issue in context. After the short documentary, we moved to in-studio dialogue with experts and a diverse studio audience. As key questions arose in the studio discussion, they were presented to the scientific sample who viewed the *Community Voice* program from their homes. They dialed-in their votes, which were then displayed both to participants in the studio and to viewers at home. Six votes were easily taken during the prime-time, hour-long program, which was viewed by more than 300,000 people in the Bay Area. With six votes, the overall views and attitudes of the Bay Area public were clearly established. (See the first 3 and a half minutes of this video clip.)[191]

The success of our pilot in 1987 just begins to demonstrate the potential for achieving a dramatic increase in the scope and depth of metropolitan-scale dialogue and consensus building. It is now entirely possible to develop non-partisan *Community Voice* organizations or ETMs that combine broadcast television with internet-based feedback from a scientifically selected sample of citizens. With these simple tools, the public can get to know its collective mind with a high degree of accuracy. With regular Electronic Town Meetings, the perspectives and priorities of citizens can be swiftly brought into public view and the democratic process uplifted to a new level of engagement and function.

The value and purpose of *Community Voice* organizations is not to micromanage government through direct democracy; rather, it is for citizens to discover their widely shared concerns and priorities that can guide their representatives in government. In my view, it is not the purpose of *Community Voice* organizations to become directly involved in complex policy decisions; instead, it is to enable citizens to express their overall views that can guide policymaking. Involving citizens in choosing our path into the future will not guarantee that the "right" choices will always be made, but it will guarantee that citizens will be involved and invested in those choices. Rather than feeling cynical and powerless, citizens will feel engaged and responsible for our collective future.

Major metropolitan areas around the world are the natural scale for organizing this new level of citizen dialogue and consensus building. Leadership in one community could inspire other communities to create their own *Community Voice* organizations, and an entirely new layer of sustained and meaningful dialogue could quickly sweep across countries and around the Earth. Citizens could voice their views, propose and debate solutions, and help break through gridlock.

In getting *Community Voice* organizations underway, no factor will have a greater impact on the design, character, and implementation of electronic town meetings than who sponsors them. Consider three major possibilities:

- First, if ETMs (Electronic Town Meetings) are sponsored by commercial TV stations, they will be designed to sell advertising and entertain audiences—not to inform citizens and involve the public in choosing its future.

- Second, if ETMs are sponsored by local, state, or national governments, they would likely be used as public relations tools, rather than as an authentic forum for open community dialogue.

- Third, ETMs sponsored by issue-oriented organizations or institutions representing a particular ethnic, racial, or gender group would likely focus on the concerns of that group.

A critical conclusion emerges: *an independent, non-partisan* Community Voice *organization is needed to act on behalf of all citizens* as the sponsor of electronic town meetings. Once *Community Voice* organizations are established and operating in major metropolitan areas, it would be entirely practical to join together to create regional ETMs; for example, seacoast cities could join in a common effort to respond to sea-level rise. Once regional ETMs are underway and securely anchored with trusted communication, the next step would be to create national dialogues for the future we want. Looking beyond regional and national ETMs, we already have the technological capacity to create global ETMs with

an *Earth-Voice* system *that could supercharge humanity's uplift at a planetary scale.* Earth Voice is practical and doable:
- *Television*: Already, between three- and four-billion people watch the Olympics on television globally.[192] A majority of Earth citizens have access to TV sets within reach of a TV signal.[193]
- *Internet*: In 2021, roughly 65 percent of the global population had internet access.[194] Internet access is expected to reach 75 percent of the global community by the end of this decade.[195]

Although we are slow to recognize the immense power of a non-partisan *Earth Voice* movement, we already possess tools with the astonishing power to enable us to begin communicating our way into a workable and purposeful future.

> *The next great superpower will not be a nation or even a collection of nations; rather, it will be the billions of ordinary citizens who encircle the Earth and who call, with a collective voice, for unprecedented cooperation and creative action to care for our endangered Earth and for humanity to grow into a mature planetary civilization.*

A new superpower is emerging from the combined voice and conscience of the world's citizens, mobilized through a local-to-global communications revolution. When the people are more than passive recipients of information—as *witnesses* to climate disruption, intense poverty, and species extinction—but are also able to offer a collective *voice* for change, then a new and powerful force for creative transformation will be unleashed in the world. And just in time! *Never before in history have so many people been called upon to make such sweeping changes in so little time.*

Once citizens know what other citizens around the world are willing to do, and once they are settled in their own hearts and minds as to what constitutes appropriate action, then they and their representatives in government can act swiftly and with authority. Democracy has often been called the art of the possible. If we don't know how our fellow citizens think and feel about collective efforts to create a sustainable and purposeful future, then we float powerlessly in a sea of ambiguity—unable to mobilize ourselves to constructive action. A mature democracy and society require the active participation and consent of an informed public, not simply passive acquiescence. Once humanity develops the simple capacity for sustained and authentic social reflection, we will have the means to achieve a shared understanding and working consensus regarding appropriate actions for a positive future. Actions can then come quickly and voluntarily. We can mobilize ourselves purposefully, and each person can contribute his or her unique talents to building a life-affirming future. I agree with Lester Brown, president of the Worldwatch Institute, who said "The communications industry is the only instrument that has the capacity to educate on a scale that is needed, and in the time available."

Choosing Maturity

In speaking to diverse audiences around the world for the past 40 years, I have often begun by asking a simple question: "When you observe the human family and our behavior, what is your view of the overall life-stage of our species? Are we behaving like infants, adolescents, adults, or elders?" I have asked this same question to diverse business leaders in Brazil, the U.S., and Europe; to spiritual leaders in Japan and the U.S.; to women graduating as schoolteachers in India; to non-profit groups and student groups in the U.S., Canada, and Europe; to an international community of women leaders, and more. Wherever I have asked this question, the answer is immediate and overwhelming: *Roughly three-quarters say that humanity, taken as a whole, is in its adolescent stage of behavior as a species!* The most common reasons offered for this point of view are:

- Adolescents are often *rebellious* and want to prove their independence. Humanity has been rebelling against nature, trying to demonstrate our independence and superiority.

- Adolescents can be *reckless* and inclined to live without regard for the consequences of their behavior, often feeling they are immortal. The human family has been recklessly consuming natural resources as if they would last forever, polluting

the air, water, and land, and eliminating a significant part of animal and plant life on Earth.
- Adolescents are frequently concerned with outer *appearance* and with fitting-in materially. Many humans are concerned with how they express their identity and status through material possessions.
- Adolescents are inclined toward instant *gratification*. As a species, we seek short-term pleasures, largely ignoring the long-term needs of other species or our own future generations.
- Adolescents tend to gather in groups or cliques, and often express this as *"in-vs.-out"* thinking and behavior. Much of humanity is clustered into political, socio-economic, racial, religious, and other groupings that separate us from one another, fostering an "us vs. them" mentality.

I see a hopeful possibility in these results. If we can move from our collective adolescence to early adulthood, rebellion can shift to collaboration; recklessness can grow into discernment; absorption in outward appearance can give way to attention to inner integrity; a focus on personal gratification can grow into a desire to be of service to others; and separation into cliques and in-groups can grow into concern for the well-being of a larger community.

Adolescents have important qualities we need as we mature into early adulthood: they often have huge amounts of energy and enthusiasm and, with their courage and daring, are ready to dive into life and make a difference in the world. Many adolescents have a hidden sense of greatness and feel that, if given a chance, they can accomplish remarkable things. In stepping into our early adulthood as a species we can liberate ourselves from the constraints of the past, awaken untapped energy, creativity, and courage, and work to achieve greatness that is now hidden.

Growing up is entirely natural, yet it is important to acknowledge how demanding this journey is: Maya Angelou wrote these powerful lines describing the difficulty of growing up:

> I am convinced that most people do not grow up. We find parking spaces and honor our credit cards. We marry and dare to have children and call that growing up. I think what we do is mostly grow old. We carry an accumulation of years

in our bodies and on our faces, but generally our real selves, the children inside, are still innocent and shy as magnolias.[196]

Toni Morrison said in a commencement speech, "True adulthood is a difficult beauty, an intensely hard-won glory, which commercial forces and cultural vapidity should not be permitted to deprive you of."[197]

When I ask people, what motivated them to move from their adolescence to adulthood, common themes emerge that are instructive for humanity's initiation and great transition. People often mention:

- *A brush with death*—the death of a friend or family member awakened an understanding of our mortality and how we have a limited time on Earth to learn and grow. The threat of our extinction is a powerful motivation for moving into our early adulthood.

- *Role models* inspire adolescents to reach beyond current behaviors and explore new potentials. Current role models tend to be movie stars, sports stars, and popular musicians. However, these role models tend to encourage adolescent behaviors rather than draw us into early maturity.

- Pushed to take *responsibility for the well-being of others*—for example, caring for a sibling, an aging parent, a sick friend, or taking an extra job to earn money for the family. Now we are being pushed beyond ourselves to take responsibility for the well-being of the Earth.

- Pushed to take *a "hard look in the mirror"*—seeing how we live in adolescent ways, such as prioritizing consumption over service. The internet and television give us reflective feedback and a penetrating view of ourselves. We can see more clearly the consequences of our behavior and the need to step up to a higher level of maturity.

If the human community is still generally in adolescence, it explains much of our current behavior and suggests how we could behave differently if we move collectively into early adulthood:

- **Maturing adults tend to give priority to others before themselves**. With greater maturity, adults are able to look beyond self-centered wants and desires and, instead, consider how they can serve the well-being of others and the Earth. Rather than being self-absorbed, adults can be selfless and make sacrifices for others without feeling resentful. A mature person and society can find joy in the success of others and derive satisfaction from sharing their good fortune with others.

- **Maturing adults tend to keep long-term commitments and choose delayed gratification**. If we are to make a commitment to the well-being of future generations and pull back from overconsuming the Earth, then a higher level of maturity is vital. Beyond token generosity, the global society and economy need to be reconfigured for equity and common good. This is truly an undertaking for mature adults.

- **Maturing adults tend to have a greater sense of humility**. Adults are more unpretentious and feel less concerned with the need to prove themselves to others; instead, they tend to choose more modest ways of being and living. With greater maturity comes greater concern for fairness and the equal rights of others.

- **Maturing adults tend to be more self-accepting and accepting of others**. A mature person or society has been seasoned by life experience and tends to realize we are here for more than pleasure seeking—we are here to learn, grow, and give to the well-being of others. In maturity, we accept our humanity and have greater compassion for ourselves and others.

- **Maturing adults tend to talk less and listen more**. A mature person will tend to listen for understanding, rather than listen for opportunities to interrupt and argue for their point of view. In our times of growing tensions and conflicts, we need to listen deeply, especially to younger and marginalized populations. Listening and learning go together as invaluable skills for a world in great transition.

- **Maturing adults tend to clean up after themselves**. Adults don't expect others to clean up the messes they make.

Instead of waiting for others to handle things, adults take charge of their own lives.

- **Maturing adults recognize that failure and missteps are part growth**. We will not always live from our highest ideals, morals, or qualities. Mature people will acknowledge when they are out of alignment with their values and commitments, and then integrate what they have learned so they can do better.
- **Maturing adults are aware that each of us has blind spots**. Maturing involves recognizing that our points of view can limit how we see and understand ourselves, others, and the larger world. Maturing means recognizing our own biases and limitations and, with a measure of humility, developing empathy for other people's perspectives and points of view.

These practical and significant changes, taken together, could bring a tremendous uplift to the human journey. They reveal that one of the most significant changes needed is for humanity to recognize how we are deeply embedded in an interconnected web of relationships. Human survival now depends on humanity waking up and taking our place within the web of life, becoming responsible co-creators with the rest of life, and living with conscious regard, reverence, and care for the well-being of all life.

Choosing Reconciliation

The many divisions in our world absorb a tremendous amount of time and energy that, if healed, could free up the energy and attention needed to promote and create a workable and purposeful world. Conflicts, agitation, pushback, antagonism, etc. occupy personal and public attention and distract us from coming together to find a higher common ground for coping with the existential crisis to our collective future. Truly, we face the possibility of our functional extinction as a species and without healing these divisions our efforts for a sustainable and regenerative future will fall short.

Injustice and inequities flourish in the darkness of inattention. Being exposed to the healing light of public awareness creates a new consciousness among everyone involved. With the revolution in communication, the world becomes transparent to itself. Increasingly, the media bring injustice, oppression, and violence, into the court of public attention and opinion. In our communications-rich and tightly interdependent world, it will be difficult for old forms of repression and violence to continue without world public opinion turning back on the oppressors.

As our capacity for collective consciousness awakens, the deep psychic wounds that have festered throughout humanity's history will come to the surface. We will hear the voices that have been unacknowledged and the pain that has been unexpressed. Professor Christopher Bache explains:

> "The floor of the collective unconscious appears to be rising. As it does, it is bringing with it the psychic sludge of history. The first step toward realization is always purification. The karmic residue of the choices made by countless generations of half-conscious human beings is rising into our individual and collective awareness as we confront *en masse* the legacy of our past."[198]

It might seem unwise to bring the dark side of humanity's past into the light of day, but unless we do, this unresolved pain will forever pull at the underside of our consciousness and diminish our future potentials. Fortunately, the compassionate clarity of reflective consciousness provides the psychological space for healing to occur.

Being heard is the first step to being healed. When we feel met and heard with the active listening of others, we open more fully to our sorrows as well as to those of others. By acknowledging and listening to the stories of those who have suffered, we build a foundation of compassion to aid the process of healing. Collective listening to stories of humanity's wounding is vital for society's healing. Healing means that we publicly acknowledge and mourn legitimate grievances and seek just and realistic remedies.

In its simplest terms, cultural healing means overcoming our deep separations—from one another, from the Earth, and from the living cosmos. Healing occurs when we realize that the life-force uniting us is deeper than the differences that divide us. With conscious cultural healing, the human family can advance beyond chronic ethnic conflict, racial oppression, economic injustice, gender discrimination, and the other inhumanities that divide us. If we can bear witness to the reservoir of unresolved pain accumulated through history, we will release an enormous store of pent-up creativity and energy. We can realize a tremendous evolutionary uplift with the release of humanity's collective energy in service of building a positive and nurturing future. What a remarkable species-project this could make. As humanity's inner world of experience consciously engages the outer world of action, we can begin our common work of building a sustainable, satisfying, and soulful species-civilization.

All people share the common ocean of consciousness. Irrespective of differences in gender, race, wealth, religion, and so on, we all participate in the deep ecology of consciousness, and this provides a common ground for meeting, for mutual understanding, and for reconciliation. Reconciliation does not mean that past injustices and grievances are erased; rather, by being consciously acknowledged and coupled with sincere efforts for restoration, they no longer stand in the way of our collective progress. When injustices are consciously recognized along with public apology and reparation, it releases both parties from the need to continue the process of blaming and feeling resentful, instead both can focus on restorative and cooperative actions for building a constructive future. The Earth community faces a stark choice for the future—do we:

- **pull together** as a human community, accepting all the *sacrifices* that will be involved, or
- **pull apart** as human sub-groups, enduring all of the *violence* that will inevitably result?

With reconciliation and pulling together, we humans can truly realize astonishing achievements. Genuine uplift can come from healing the wounds of division and coming together in common efforts as a species. This is not fantasy, but the clear reality of our current world situation. We are so divided in so many ways that working together in a common effort seems nearly impossible. However, the fiery passage through our time of great initiation can burn through the many barriers that now divide us from wholeness and collective effort as a species

If the Earth community chooses to pull together and collaborate for the well-being of all, a cascade of actions and innovations can follow quickly from the clarity of our unified social will. However, if the social will of the people is not awakened on behalf of our *collective* well-being, but remains deeply divided, then we seem likely to turn toward either the seeming safety of authoritarianism or fragment into countless sub-groups, as unresolved wounds and divisions continue, generating ever-deeper separation and increased violence.

Only together can we realize a great transition to planetary community. Transition is a team effort—all hands, on deck! A team effort is impossible if we are deeply divided as a human community. The world is awash with racial and gender discrimination, genocide, religious wars, oppression of ethnic minorities, and the extinction of other species. Some of these tragedies have grown and festered over thousands of years, and this makes pulling together in common effort extremely difficult. Nonetheless without deep and authentic reconciliation across these and other barriers, humanity will remain separated and mistrustful—and our collective future will be gravely imperiled.[199] As difficult and uncomfortable as this process might be, conscious reconciliation that includes truth telling, public apology, and meaningful reparation is a vital part of our collective healing—essential if humanity is to move forward together on our journey.

A world divided against itself is a recipe for global collapse and humanity's functional extinction. We can acknowledge the wisdom of Dr. Martin Luther King, Jr.: "We must learn to live together as brothers or perish together as fools."[200] In the words of South African anti-apartheid activist Alan Paton, "It is not 'forgive and forget' as if nothing wrong had ever happened, but 'forgive and go forward,' building on the mistakes of the past and the energy generated by reconciliation to create a new future."[201]

Although we can see the broad outlines of a sustainable future, the human family is far from being ready to work together. To come together, the Earth family must engage in a process of authentic reconciliation across a number of areas:

- **Gender, racial, sexual, and ethnic reconciliation**—Discrimination profoundly divides humanity against itself. To work together for our common future, we must build a global culture of mutual respect that enables us to work together as equals. This does not mean we will ignore gender, racial, sexual, and ethnic differences; rather we learn to respect and include differences, and then work to transform oppressive structures and systems. We move beyond limiting judgments of others and weave a new culture of respect, inclusivity, and fairness.

- **Generational reconciliation**—Sustainable development has been described as that which meets the needs of the present without compromising the ability of future generations to meet their needs.[202] Because many industrial nations are using up vital nonrenewable resources in the short run, the options available for future generations to meet their needs will be severely limited. To pull together, we must reconcile ourselves across generations. For example, adults can support youth by listening to their needs, shining a light on youth movements and concerns, and hearing how the lifestyles of the current generation have helped to create the climate crisis.

- **Economic reconciliation**—Enormous disparities exist between rich and poor. Reconciliation requires narrowing these differences and establishing a global minimum standard for economic well-being that supports people to realize their potentials. Yale professor, Narasimha Rao states, "reducing inequality—within countries and between them—would improve our ability to mitigate some of the worst effects of climate change, and provide for a more stable climate future . . . climate change, at its most essential, is a justice issue."[203] United Nations research reveals that global inequality is often more about disparities in opportunity than disparities in income.[204] Perhaps the deepest change will be to disconnect how personal value is associated with one's position in a hierarchy of wealth or social class.

- **Ecological reconciliation**—Living in sacred harmony with the Earth's biosphere is essential if we are to survive and evolve as a species. Restoration of the biosphere is vital because our common future depends on the presence of a broad diversity of plants and animals. To move from indifference and exploitation to reverential stewardship will require reconciliation with the larger community of all Earth life, and honoring those who have preserved cultures of sacred reciprocity with all life. Consumer cultures place the material wants of a few above the needs of the overall Earth community, and this has led to ecological disasters. We humans are an inseparable part of the Earth, and what happens to the Earth happens to us.

- **Religious reconciliation**—Religious intolerance has produced some of the bloodiest wars in history. Vital to humanity's future is reconciliation among the world's spiritual traditions—for example, Catholics and Protestants in Northern Ireland, Arabs and Jews in the Middle East, Muslims and Hindus in India. As the world's religious and spiritual traditions become more accessible through the internet and social media, we can discover the core insights of each tradition and see each as a different facet in the common jewel of human spiritual wisdom.

Many of these divisions are starkly evident in our world and, with climate disruption, will disproportionately impact women and the world's poor most profoundly. Here is a compelling summary from a recent Oxfam briefing:

> "Within countries, it is often the poorest communities—and particularly women—who are most vulnerable. Poor communities tend to live in poorly built houses on marginal land that is more at risk from extreme weather such as storms or floods. They often live in areas with poor infrastructure, making it difficult to access essential services such as healthcare or education in the aftermath of an emergency. They are unlikely to have insurance or savings to help them rebuild their lives after a disaster. And many depend on farming or fishing—activities that are particularly vulnerable to more extreme and erratic weather. With the frequency and intensity of climate-related hazards increasing, the ability of people living in poverty to withstand shocks is gradually being eroded. Each disaster is leading them in a downward spiral of deeper poverty and hunger, and eventually displacement. . . . When forced to leave home, women and

children are particularly vulnerable to violence and abuse.... Displaced children are often denied an education, locking them in an inter-generational cycle of poverty."[205]

The emergence of a "living-universe" paradigm reawakens a deep feminine perspective that honors the unity of life.[206] From at least 50,000 years ago until roughly 6,000 years ago, an "Earth Goddess" perspective guided the relationship of humans with the larger world.[207] The feminine archetype recognized and honored the aliveness and regenerative powers of nature and the fertility of life. Then, roughly 6,000 years ago, with the rise of city-states, more differentiated classes (priests, warriors, merchants), and more complex cultures, a masculine mindset and "Sky-God" spirituality became dominant and supported the development of human society organized into larger-scale structures and institutions. A masculine, patriarchal mindset has grown and developed over thousands of years and has encouraged the growing individuation, differentiation, and empowerment of people. It also supported humanity's growing separation from, and exploitation of, nature that has led to our current ecological crisis. A "Cosmic Goddess" perspective, by contrast, regards the generative and sustaining nature of the universe more from a feminine point of view. Overcoming thousands of years of separation through deep reconciliation that honors the sacred feminine and its affirmation of the unity of life is vital if we are to rise beyond the divisions of the past.

Great personal and social maturity are required to recognize and remedy injustices and injuries, so the human family can work together for our common well-being. Bringing legitimate grievances into public awareness, mourning the mistakes of the past, taking responsibility for them, and then seeking just and realistic remedies—these difficult actions are at the heart of the era of reconciliation.

> *We require unprecedented communication*
> *to discover our common humanity*
> *from a mindset of uncommon humility.*

With reconciliation and restoration, social energy that was previously locked up in oppression and injustice can be freed up and become available for productive working relationships.

The process of reconciliation is complex and involves three major steps: the injured must be heard publicly, the wrongdoers must apologize publicly and take responsibility for the impacts of their actions, and then they must provide restitution or reparations that make amends for the past, and provide a foundation of greater wholeness for all to move into the future together.

Being heard is the first step in being healed. By listening to and acknowledging the stories of those who have suffered, we begin the process of healing. Our collective listening to the wounds of humanity's psyche and soul is vital to our collective healing. Listening does not mean forgetting; instead, it means bringing the wounds of division into collective awareness and remembering these as we seek ways to move into the future.

Archbishop Desmond Tutu knew more about the process of reconciliation than most. He was the chairman of the Truth and Reconciliation Commission (TRC), established to investigate crimes committed during the apartheid era in South Africa from 1960 to 1994. When apartheid ended, South Africa's black majority had to choose among three different ways to seek justice and live together with the country's white minority. They could choose justice based on *retribution*—an eye for an eye; or justice based on *forgetting*—don't think about the past, just move forward into the future; or justice based on *restoration*—granting amnesty in exchange for truth. Archbishop Tutu explained their choice:

> "We believe in restorative justice. In South Africa, we are trying to find our way toward healing and the restoration of harmony within our communities. If retributive justice is all you seek through the letter of the law, you are history. You will never know stability. You need something beyond reprisal. You need forgiveness."[208]

A second step in being healed is for the wrongdoer to offer a sincere public apology. Here are examples of important public apologies:[209]

- In 1988, an act of Congress apologized "on behalf of the people of the United States" for the internment of Japanese Americans during World War II.

- In 1996, German officials apologized for the invasion of Czechoslovakia in 1938 and established a fund for the reparation of Czech victims of Nazi abuses.
- In 1998, the Japanese prime minister expressed "deep remorse" for Japan's treatment of British prisoners during World War II.
- In 2008, the U.S. Congress formally apologized for the country's "original sin"—its treatment of African Americans during the era of slavery and subsequent laws that discriminated against blacks as second-class citizens in U.S. society.

Another powerful example of a public apology and social healing is an attempt to heal the relationship between the Aboriginal people and the European settlers in Australia. In 1998, Australia commemorated its first "Sorry Day" to express regret and shared grief about a tragic episode in Australian history—the organized removal of Aboriginal children from their families on the basis of race.

Through much of the 20th century, Aboriginal children were forcibly removed from their families with the aim of assimilating them into Western culture.[210] "Sorry Day" marks a way for Australians to come to terms with their history and to remember together, as a way to build a future on a foundation of mutual respect. Indigenous council member Patricia Thompson stated, "What we want is recognition, understanding, respect, and tolerance—of each other, by each other, for each other." In cities, towns, and rural centers, in schools and churches, people stop their everyday activities to acknowledge this injustice. In addition, hundreds of thousands of Australians have signed "Sorry Books." An essential requirement for reconciliation is conscious requests for forgiveness as well as remembering.

The third step in reconciliation is restitution or the payment of reparations. Archbishop Desmond Tutu explained the role of restitution when he said that reconciliation involves more than the recognition and remembering of injustice: "If you steal my pen and say 'I'm sorry' without returning the pen, your apology means nothing."[211] Restitution is also needed. Apologies create a truthful record. Restitution creates a new record. The purpose of reparation is to repair the material conditions of a group and to restore balance or equality of power and material opportunity.[212]

With authentic reconciliation—that includes listening, remembering, apologizing, and restoring—the divisions and suffering of the past do not need to stand in the way of future harmony. This is not as simple as providing money or land or policies designed to remove inequities. The deep wounding of the oppressed also manifests as generational trauma that no amount of money will erase. True reparation must provide for healing and wholeness.

As difficult and uncomfortable as this process will be, this is a vital stage in our collective healing that can bring a tremendous uplift to humanity for moving forward on our common journey. Just as a rising tide lifts all boats, so too can a rising level of global communication lift all injustices into the healing light of public awareness. Our ability to communicate with ourselves as a planetary species about these painful wounds will be critical for realizing the uplift of reconciliation.

Choosing Community

The issue of "choosing Earth" raises a further question of whether we feel a sense of belonging on the Earth? Do we feel at home here—where "home" is not only a physical place, but also a feeling in our body, heart, and soul? Does our physical home connect us with a local community that, in turn, connects us with the Earth? The home and community we inhabit carry an invisible language and feeling that is communicated in its physical structure. Architect Christopher Alexander writes about the "pattern language" communicated by the homes, communities, and cities we inhabit.

> "A pattern language expresses the deeper wisdom of what brings aliveness within the life of our community. Aliveness is a term for "the quality that has no name": a sense of wholeness, spirit, or grace, that while of varying form, is precise and verifiable in our direct experience."[213]

The qualities of aliveness expressed in the physical patterns of our homes and communities, communicate a message that can be silent to our ears, but loud to our intuitions. How can we "choose Earth" if we don't feel a part of its patterns and that we belong here?

People in more materially developed countries often seek to live in splendid isolation. In the sprawling suburbs, single-family homes are designed to be set apart from other homes, often with a fence for clear separation from neighbors. Living in contained isolation, whatever we need to support our everyday lives can be purchased at well-stocked stores or ordered online for fast delivery. There is no need to bother others or for them to bother you. Years can go by without knowing immediate neighbors.

The physical design of our homes and community creates an experience of either uplifting belonging or existential isolation. Our modern lives have often been designed for deliberate separation and this contrasts profoundly with ancient roots of tribal existence grounded in close relationships with other people, local nature, and invisible forces in the world. The African word *ubuntu* communicates the importance of community. *Ubuntu* refers to the idea that we discover ourselves through our relationships with others. *Ubuntu* is defined as knowing *"I am who I am because of who we all are."* We develop ourselves through our interactions with others. In turn, the quality of these relationships lies at the core of our lives. With *ubuntu*, we are open and available to others and feel part of a larger whole. *Ubuntu* is relationship and uplift. Isolation is alienation and downfall.

A singular, isolated, existence can work well with access to material abundance and well-functioning supply chains for purchasing food and products to support our lives. However, when supply chains break down, and money cannot buy easy access to things we need, then the quality of our relationships with others once again defines our lives.

Innovations in the physical design of communities are vital for transforming how we live on the Earth. Patterns of living that prioritize sprawling suburbs and isolated households are not well-suited for sustainability. Hyper-individualized patterns of living create formidable barriers to further innovation. Growth creates form and form limits growth. Urban growth creates a pattern of living—such as a sprawling suburb—and once these physical forms are anchored in the land, they limit the ability to create new patterns for living.

A transforming world requires new configurations of living better adapted to a rapidly changing ecology, society, and economy. In turn, a spectrum of innovation is beginning to grow from local to global levels:

- **Pocket Neighborhoods** generally consist of a few homes linked together to promote a close-knit sense of community and neighborliness with an increased level of uplifting connection.

  Pocket neighborhoods are generally clustered groups of neighboring houses or apartments gathered around a shared open space—a garden courtyard, a pedestrian street, a series of joined backyards, or a reclaimed alley—all of which have a clear sense of territory and shared stewardship. They can be in urban, suburban, or rural areas. A pocket neighborhood is *not* the wider neighborhood of several hundred households and network of streets, but a realm of a dozen or so neighbors who interact on a daily basis around a shared local commons—a kind of secluded neighborhood within a neighborhood.

- **Ecovillages** are either designed freshly or, more commonly, retrofitted to provide an integrated way of life with roughly a hundred or so people. Ecovillages are intentional communities, united by shared values and with the goal of becoming more socially, culturally, economically, and ecologically sustainable. Usually, they are locally owned and governed by participatory processes. A regular feature of many of ecovillages or co-housing communities is a common house for meetings, celebrations, and regular meals together; an organic community garden; a recycling and composting area; a renewable energy micro-grid; a bit of open space for community gatherings; perhaps a play space and conversation space for teenagers; and a workshop with tools for arts, crafts, and repair.

  Ecovillages can include a micro-economy where community members trade hours to create a local economy, offering services such as healthcare, childcare, elder care, gardening, education, green building, conflict resolution, internet and electronic support, food preparation, and other skills that provide fulfilling connection and contribution to the community. The scale is small enough for everyone to know one another and yet large enough to support a micro-economy with meaningful work roles for many. Ecovillages have the culture and cohesiveness of a small town and the sophistication of a city, as nearly everyone is connected to the world with the internet and other electronic tools for communication. Ecovillages encourage unique expressions of sustainability as they foster simplicity of living, raise healthy children, celebrate life in community with others, and seek to honor the Earth and future generations. The flowering of diverse ecovillages can bring powerful uplift to our lives.[214]

- **Transition Towns** bring together pocket neighborhoods and ecovillages into a town of several thousand people. They generally support grassroots projects that aim to increase local self-sufficiency and reduce the harmful effects of climate change and economic instability. The "Transition Network," founded in 2006, has inspired the creation of transition town initiatives around the world.[215]

- **Sustainable Cities** seek to aggregate pocket neighborhoods, ecovillages, and transition towns into a larger system of sustainable and ecological living. A sustainable city is modeled on the self-sustaining resilient structure of natural ecosystems. An ecocity seeks to provide a healthy life for its inhabitants without consuming more renewable resources than it produces, without producing more waste than it can assimilate, and without being toxic to itself or neighboring ecosystems.[216] Inhabitants tend to choose ecological ways of living that embody principles of fairness, justice, and equity.

- **Eco-Civilizations** take the lessons learned at smaller scales and extend them to nations, clusters of nations, and the entire Earth community. Eco-civilizations respond to global climate disruption and social injustices with alternative approaches to living based on ecological principles. An ecological civilization moves toward a regenerative future with a synthesis of economic, educational, political, agricultural, and social designs for sustainable living.[217]

A nested spectrum of innovation in housing, economic activity, and ecological ways of living illustrates how we are beginning to reconfigure our local lives to adapt to new global realities. The urgency of shifting to a zero-carbon economy is pushing humanity away from an "ego economy" that is devastating the Earth, toward an "aliveness economy" that uplifts our relationship with the Earth.

In our rapidly transforming world, designs for adapting our lives for uplifting forms of ecological living are emerging across a wide spectrum—from the smallest scale of pocket neighborhoods to the largest scale of entire eco-civilizations. As the century advances, millions of experiments in innovative forms of regenerative living will develop. Alternative communities of every imaginable design will adapt to local conditions and provide islands of sustainability, security, and mutual support. However, as a note of caution, the strength of local ecovillages and communities could become a weakness if they are seen primarily as isolated havens of safety to weather the storms of transition. *Lifeboats won't save us when the entire Earth is sinking and becoming inhospitable to life.* It is vital for the cohesion that develops in local collaborations to reach more widely and provide the social glue for holding together larger networks. Synergies among pocket neighborhoods and local ecovillages need to move up the scale to transition towns and sustainable cities, and finally to the scale of the world as an eco-civilization. These synergies create powerful uplift along the full spectrum of innovation.

Choosing Simplicity

The magnitude and speed of climate disruption now underway is astonishing and will require dramatic changes in how we live on Earth. For the past few hundred years, consumer-oriented societies have exploited the global resources for the benefit of a fraction of humanity. The goal of this approach has been to find happiness through consumption and to satisfy our material *wants* without conscious regard for the *needs* of a livable Earth. This self-serving approach brings ruin to the Earth and humanity's future. Instead of asking what we humans *want* (what we desire, crave, or hunger for), we are being called to respond to a far more important question: what does the overall ecology of life *need* (what is essential, basic, necessary) to build a regenerative future for the Earth? To live sustainably on the Earth, we need to choose ways of living that match our consumption with the regenerative capacities of the Earth and the needs of the rest of life with whom we share the biosphere. Instead of a wealthy minority pulling humanity down, a generous majority can live with moderation and kindness and bring tremendous uplift to living on Earth.

A study of what is required for "Life Beyond Growth" found that "a country like Japan would have to cut its consumption of resources and environmental impact by (very roughly speaking) more than 50 percent, while the United States would need to reduce by a factor of 75 percent."[218] Therefore, when we ask, "What can we do to support the ecology of life?" the first powerful action we can take is to bring our personal lives into alignment with the regenerative needs of the Earth. In addition, the affluent minority needs to recognize that an impoverished majority lives at the margins of material existence and, for them, simplicity of living is involuntary—they have few options and little choice in their daily struggles for survival.

Although simplicity is intensely relevant for building a workable world, this approach to living is not a new idea. Simplicity has deep roots in history and finds expression in all of the world's wisdom traditions. More than two thousand years ago, in the same historical period that Christians were saying "Give me neither poverty nor wealth," (Proverbs 30:8), Lao Tzu, the founder of Taoism declared, "I have just three things to teach: simplicity, patience, compassion. These three are your greatest treasures"; Plato and Aristotle proclaimed the importance of the "golden mean"—a path through life with neither excess nor deficit; and Buddhists encouraged a "middle way" between poverty and mindless accumulation. Clearly, the wisdom of simplicity is not a new revelation.[219] What is new is the reality of humanity pressing against limits to material growth and recognizing the importance of building a new relationship with the material aspects of life.

Simplicity is not opposed to consumption of resources; instead, it places material consumption in a larger context. Simplicity does not encourage turning away from material progress; to the contrary, an advancing relationship with the material side of life lies at the heart a maturing civilization. Arnold Toynbee—a renowned historian who invested a lifetime in studying the rise and fall of civilizations throughout the world—summarized the essence of a civilization's growth in what he called *The Law of Progressive Simplification*.[220] He wrote that a civilization's progress was not to be measured in its conquest of land and people; instead, the true measure of growth is a civilization's ability to transfer increasing amounts of energy and attention from the material side of life to the non-material side—areas such as personal growth, family relationships, time with nature, psychological maturity, spiritual exploration, cultural and artistic expression, and strengthening democracy and citizenship.

Recall that modern physics recognizes that 96 percent of the known universe is invisible and non-material. The material aspect (including galaxies, stars, and planets and biological being) consists of only about four percent of the known universe. If we apply these proportions to our lives, then it is fitting to give greater attention to the invisible aspects that are often ignored and that represent the very aspects that Toynbee describes as expressing our progress as a civilization.

Toynbee also coined the word "etherialization" to describe the process whereby humans learn to accomplish the same, or even greater, results using less time, material resources, and energy. Buckminster Fuller called this process "ephemeralization," although his emphasis was on realizing greater material performance for less time, weight, and energy invested. Drawing from the insights of Toynbee and Fuller, we can redefine progress as a two-fold process involving the simultaneous refinement of both the material and non-material sides of life.

> *With progressive simplification,*
> *the material side of life grows lighter,*
> *less burdensome, more easeful, elegant and effortless*
> *and, at the same time, the non-material side of life*
> *becomes more vital, expressive, and artistic.*

Simplicity involves the co-evolution of both inner and outer aspects of life. Simplicity does not negate the material side of life, rather it calls forth a new partnership where the material and the non-material aspects of life co-evolve in concert with one another.

Outer aspects include the basics such as housing, transportation, food production, and energy generation. Inner aspects include learning the skills of touching the world ever more lightly and lovingly—ourselves, our relationships, our work, and our passage through life. By refining both outer and inner aspects of life (outward simplicity combined with inner richness) we can foster genuine progress and build a sustainable *and* meaningful world for billions of people without devastating the ecology of the Earth.

An ethic of moderation and "enough" will grow in importance as global communications reveal vast inequities in material well-being. Economic justice does not require replicating the industrial-era mode of life globally; instead it means every person has a right to a fair share of the world's wealth, adequate to ensure a "decent" standard of living—enough food, shelter, education, and healthcare sufficient for a reasonable standard of human decency.[221] Given intelligent designs for living lightly and simply, a decent standard and manner of living could vary significantly depending on local customs, ecology, resources, and climate.

To accomplish a great transition within a few decades requires that we invent new approaches to living that transform every facet of life—the work we do, the communities and homes in which we live, the food we eat, the transportation we use, the clothes we wear, the symbols of status that shape our consumption patterns, and so on. We can call this way of living "voluntary simplicity" or "conscious simplicity" or "ecological living."[222] However described, we need more than a change in our style of life.

A change in *style* implies a superficial or exterior change—a new fad, craze, or fashion. We require a far deeper change in our *way* of life, one that recognizes the Earth is our home and must be maintained for the long-range future. Ecological living begins with the understanding that we all live in mutual contingency and that we also create safety, comfort, and compassion in our lives together.

An ecologically conscious economy will shift its emphasis from sheer physical expansion to more qualitative growth of greater richness, depth, and connection. Products will be designed with increasing efficiency (doing ever more with ever less), while simultaneously increasing their beauty, strength, and ecological integrity.

Voluntary simplicity does not encourage a life of poverty, deficiency, and deprivation, when living can be transformed through intelligent design into elegant simplicity.[223] The level of satisfaction and beauty in living can be increased while lowering the quantity of resources consumed and the amount of pollution produced.

How can we awaken a new regard for living simply in a world so focused on material consumption? To make a turn toward simplicity and sustainability, it is helpful to recall the paradigm of aliveness and how, for tens of thousands of years, our ancestors were aware of living within a subtle ecology of aliveness. That awareness has been temporarily replaced by the view that our universe consists of primarily of dead matter and empty space, without purpose or meaning. Recall the logic of the two paradigms considered earlier:

- If the universe is viewed as dead at its foundations, then it is natural to exploit the Earth and use it up;
- If the universe is viewed as alive at its foundations, then it is natural to cherish the Earth and care for it.

How can we shift to a mindset of regenerative living when so much of the world currently lives within a mindset of exploitive living? A perceptive quotation from Antoine de Saint-Exupery suggests a way: "If you want to build a ship, don't drum up people to collect wood and don't assign them tasks and work, but rather teach them to long for the endless immensity of the sea." This wisdom suggests that if we want to build a regenerative world, then don't drum up people to collect materials and assign tasks for them to do; instead, *teach people to long for the endless immensity of our living universe and their unique ways of participation within it*. Awakening a longing for living in the unbounded enormity and richness of our living universe will naturally draw out people's energy and creativity for building a regenerative and beautiful world.

If we regard aliveness as our greatest wealth, then it is only natural to choose ways of living that afford greater time and opportunity to develop the areas of our lives where we feel most alive—in nurturing relationships, caring communities, time in nature, in creative expression, and service to others. In seeing the universe as alive, we naturally shift our priorities from an ego-economy oriented toward consuming dead things to an economy oriented toward growing experiences of aliveness.

An aliveness economy seeks to touch life more lightly while generating an abundance of meaning and satisfaction. Theologian Matthew Fox has written, "Luxury living is not what living is about. *Living* is what living is about! But living takes discipline and letting go and doing with less in a culture that is overdeveloped. It takes a commitment to challenge and adventure, to sacrifice and passion."[224]

In more affluent societies, consumerism is increasingly regarded as a less rewarding life pursuit and, instead, new sources of well-being are increasingly valued.[225] A major study in the U.S. by Pew Research illustrates the growing importance of direct experience over material consumption. When asked what brings the most meaning to their lives, people replied: "spending time with family" (69%), "being outdoors" (47%), "spending time with friends" (47%), "caring for pets" (45%), and "religious faith" (36%). These are not expensive—quality time with family, friends, pets, and nature is a source of richness available to nearly all.

Further evidence that wealthier nations are ready to trade reduced levels of material consumption for higher levels of experiential riches is found in a study reported in the *Wall Street Journal*:

> "People think that experiences are only going to provide temporary happiness, but they actually provide both more happiness and more lasting value [than material consumption]. Experiences tend to meet more of our underlying psychological needs. They're often shared with other people, giving us a greater sense of connection, and they form a bigger part of our sense of identity."[226]

A shift toward "postmaterialist" values is also found in the highly regarded *World Values Survey*, which concluded that, over a period of roughly three decades (1981–2007), a "postmodern shift" in values has been occurring in a cluster of a dozen or so nations—primarily in the United States, Canada, and Northern Europe. In these societies, emphasis is shifting from economic achievement to post-materialist values that emphasize individual self-expression, subjective well-being, and quality of life.[227]

Although simplicity has a long history, we are now entering radically changing times—ecological, social, economic, and psychospiritual—and we should expect the worldly expressions of simplicity to evolve and grow in response. Simplicity is not simple. A wide diversity of expressions portrays the simple life and the most useful way of describing this approach to living is with the metaphor of a garden.

Suggesting the richness of simplicity, here are ten different flowerings of expression that I see growing in the "garden of simplicity." Although they overlap to some extent, each expression of simplicity seems sufficiently distinct to warrant a separate category. (To avoid favoritism, I have ordered them alphabetically, based on the brief name associated with each.)

1. Artistic Simplicity: Simplicity means the way we live our lives represents a work of unfolding artistry. Leonardo da Vinci said, "Simplicity is the ultimate sophistication." Gandhi said, "My life is my message." Frederic Chopin said, "Simplicity is the final achievement . . . the crowning reward of art." In this spirit, artistic simplicity refers to an understated, organic aesthetic that contrasts with the excess of consumerist lifestyles. Drawing from influences ranging from Zen to the Quakers, simplicity is a path of beauty that celebrates natural materials and clean, functional expressions.

2. Choiceful Simplicity: Simplicity means taking charge of lives that are too busy, too stressed, and too fragmented. Simplicity means choosing our unique path through life consciously, deliberately, and of our own accord. It means to live whole—to not live divided against ourselves. This path emphasizes the challenges of freedom over the comfort of consumerism. Conscious simplicity means staying focused, diving deep, and not being distracted by consumer culture. It means consciously organizing our lives so we give our "true gifts" to the world—to give the essence of ourselves. As Ralph Waldo Emerson said, "The only true gift is a portion of yourself."[228]

3. Compassionate Simplicity: Simplicity means to feel such a strong sense of kinship with others that, as Gandhi said, we "choose to live simply so that others may simply live." Compassionate simplicity means feeling a bond with the community of life and being drawn toward a path of reconciliation—especially with other species and future generations. Compassionate simplicity follows a path of cooperation and fairness, seeking a future of mutually assured development for all.

4. Ecological Simplicity: Simplicity means to choose ways of living that touch the Earth more lightly and that reduce our ecological impact. This life-path remembers our deep roots in the natural world. It encourages us to connect with nature, the seasons, and the cosmos. Natural simplicity responds to a deep reverence for the community of life and accepts that the non-human realms of plants and other animals have their dignity and rights as well. Albert Schweitzer wrote, "From naïve simplicity we arrive at more profound simplicity."

5. Economic Simplicity: Simplicity means a choice for conscious consumerism and a sharing economy. Economic simplicity recognizes that we manage our relationship with our home—the Earth—by developing fitting forms of "right livelihood." It also recognizes the deep transformation in economic activity needed to live sustainably by redesigning products and services of all kinds—from housing and energy systems to food and transportation systems.

6. Family Simplicity: Simplicity means giving priority to the lives of our children and family and not to get sidetracked by consumer society. A growing number of parents are opting out of consumerist lifestyles and seeking ways to bring life-enhancing values and experiences into the lives of their children and family.

7. Frugal Simplicity: Simplicity means cutting back on spending that does not truly serve us and practicing skillful management of our personal finances—all of which can help us achieve greater financial independence. Frugality and careful financial management bring increased financial freedom and the opportunity to more consciously choose our path through life. Living with less also decreases the impact of our consumption on the Earth and frees resources for others.

8. Political Simplicity: Simplicity means organizing our collective lives in ways that enable us to live more lightly and sustainably on the Earth, and this, in turn, involves changes in nearly every area of public life—zoning, education, transportation, and energy systems. All of these involve political choices. The politics of simplicity involves, also, media politics—because the mass media are the primary vehicles promoting mass consumerism.

9. Soulful Simplicity: Simplicity means to approach life as a meditation and to cultivate an intimate connection with all that exists. A spiritual presence infuses the world and, by living simply, we can more directly awaken to the living universe that surrounds and sustains us, moment by moment. Soulful simplicity is more concerned with consciously tasting life in its unadorned richness than with a particular standard or manner of material living. In cultivating a soulful connection with life, we tend to look beyond surface appearances and bring our interior aliveness into relationships of all kinds.

10. Uncluttered Simplicity: This means cutting back on trivial distractions, both material and non-material, and focusing on essentials—whatever those may be for each of our unique lives. As Thoreau said, "Our life is frittered away by detail. . . . Simplify, simplify." Or, as Plato wrote, "In order to seek one's own direction, one must simplify the mechanics of ordinary, everyday life."

As these approaches illustrate, the growing culture of simplicity contains a flourishing garden of expressions whose great diversity—and intertwined unity—create a resilient and hardy ecology of learning about how to live more sustainable and purposeful lives. As with other ecosystems, diversity of expressions fosters flexibility, adaptability, and resilience. Because so many different paths can lead us into the garden of simplicity, this way of life has enormous potential to grow—particularly if it is nurtured and cultivated in the mass media as a legitimate, creative, and promising path for a future beyond materialism and consumerism.

Choosing Our Future

"Start by doing what's necessary; then do what's possible; and suddenly you're doing the impossible."
—Francis of Assisi

The transition to our early adulthood as a species is the most pivotal, momentous, and far-reaching transition we humans will ever be called to make. We are closing a door to the past and awakening to a new beginning. We can draw on forces of tremendous uplift as we journey into our maturity as a species. We can ride elevating and inspiring potentials of an awakening humanity and rise into a new world and new life. Our seeming downfall is the prelude to our rise. With courage, we can catch the updraft of possibilities and rise as a human community.

Reviewing the immensely powerful and yet largely untapped potentials for uplifting into a transforming future, it is abundantly clear that we could accomplish this. We face terrible consequences if we do not rise to the opportunity of choosing a new path ahead—either the functional extinction of our species along with much of the life on Earth, or a terrifying descent into authoritarianism where many of our most precious potentials are forever surrendered. We have no time left for denial or delay. The time of reckoning is upon us. Although the hour is alarmingly late, the potential for rising into a transforming pathway is still present. Uplift is neither fantasy nor make-believe hope. The forces for uplift call us to move together through a difficult transition as a species that will profoundly change our sense of who we are and the journey we are on. Uplift calls forth a new humanity; the call and the potentials are real, actual, genuine. Let's summarize them to emphasize their authentic promise. Uplift involves:

1. Choosing to live from our direct experience of *aliveness* offers a trustworthy guide for learning to live in a living universe.

2. Choosing *reflective consciousness* brings a mature regard for looking at life and the choices for our journey ahead.

3. Choosing to mobilize our potentials for local-to-global *communication* brings our collective voices into a shared conversation for the future.

4. Choosing to grow into our early adulthood awakens greater *maturity* with conscious regard for the well-being of life.

5. Choosing *reconciliation* and consciously seeking to heal the wounds of history allows us to move forward with common effort.

6. Choosing to pull together with a feeling for local-to-global *community* brings a welcoming sense of home for our journey ahead.

7. Choosing *simplicity* as a way of life that is outwardly more simple and inwardly more rich brings realism and balance into our approach for living.

When these seven factors come together in a mutually supportive approach to living, they bring the potential for soaring uplift on the human journey. If we collectively choose *aliveness, consciousness, communication, maturity, reconciliation, community,* and *simplicity,* we can awaken a nearly unstoppable force for moving through our collective initiation as a species and into a welcoming future. If we can imagine how we can move through this rite of passage, then it is our responsibility to try. What is possible becomes essential. What is feasible becomes vital. What is practical becomes critical.

A transformed humanity—and Earth—can emerge by awakening these uplifting capacities. The power of these potentials is far greater than we can imagine. With trust, we can consciously live into them and, in the process, more deeply discover ourselves. Roger Walsh, psychiatrist, and life-long meditator and teacher, writes, "We go deeper into ourselves in order to go more effectively out into the world, and we go out into the world in order to deepen into ourselves."[229] We are being called to an uplifting journey into which we can whole-heartedly invest our unique and precious lives.

Acknowledgements

This book has been a team effort and I want to express my enormous gratitude to all those who have helped bring it to life. Research, writing, and outreach for *Choosing Earth* has been supported by the courageous and generous funding of the Roger and Brenda Gibson Family Foundation. Roger and Brenda have been key allies and soulful friends in this intensely demanding work. I would not have been able to complete this book—the culmination of a lifetime of research, writing, and learning—without their support, friendship, and trust in me. They have not only supported this book, but also the larger project and learning resources that go with it. I am profoundly grateful for their partnership in helping birth and bring this work into the world.

My great appreciation also goes to Fred and Elaine LeDrew for their annual contributions to this pioneering work. Their modest donations have been large as a message of support and love. I am hugely grateful to other contributors who have made vital contributions to this project: Bill Melton and Mei Xu, Lynnaea Lumbard, Vivienne Verdon-Roe, The Betsy Gordon Foundation, Scott Elrod, Ben Elgin, Justyn LeDrew, Barbara and Dan Easterlin, Carol Normandi, Lyra Mayfield and Charlie Stein, Arthur Benz, Lorraine Brignall, Frank Phoenix, Erik Schten, Scott Wirth, Sandra LeDrew, Charles Gibbs, Marianne Rowe, Kathy Kelly, and Darlene Goetzman. Roger Walsh has contributed in a number of ways to this project and I am very grateful for his support and friendship.

My partner and wife, Coleen LeDrew Elgin, has been a key collaborator in all the facets of this creative enterprise. She produced and directed the insightful, integrative, and highly regarded documentary that is a companion for the project: *Facing Adversity: Choosing Earth, Choosing Life*. Coleen also co-taught with me, led curriculum development for the courses that accompany this book, and has provided important leadership as co-director of the project. Overall, this initiative could not have taken root without Coleen's tireless and skillful efforts for which I am tremendously grateful.

I greatly appreciate the skillful editing of Christian de Quincey who brought his keen eye to correcting and smoothing the flow of writing in this extensively revised edition. I am also enormously grateful for the thoughtful feedback and discerning suggestions the following persons have offered to this book: Coleen LeDrew Elgin, Laura Loescher, Sandy Wiggins, Roger Gibson, Brenda Gibson, David Christel, Ben Elgin, Scott Elrod, Marga Laube, Bill Melton, Eden Trenor, and Liz Moyer.

I am grateful to those who have been a part of the facilitation and teaching team for courses that have been a companion to this book: Carol Normandi, Barbara Easterlin, Sandy Wiggins, Marianne Rowe, Jim Normandi, Kathy Kelly, Diana Badger, and James Wiegel.

Birgit Wick has brought her artistry and skills into the design and layout of this book as well as to other materials in this project. She brought a caring spirit and meticulous attention to all phases of the book design and layout. For all of this I am very grateful. My thanks to Karen Preuss who photographed the hands for the cover. I appreciate Isabel Elgin for lending a hand for the cover image.

A Personal Journey

Born in 1943, I grew up on a family farm a couple miles outside a small town in southern Idaho. We lived close to the land, the passing of seasons, animals, and one another. I did not see a television until I was eleven. So, without a regular newspaper, and with only three local radio stations (that played mostly country music and commercials), my regular companions were our farm animals (dogs, cats, chickens, pigs, a horse, and a cow), the surrounding land, and the neighbors in nearby farms. I was curious as a young person and loved to read. I also enjoyed building things alongside my dad in his well-equipped wood-working shop where he would build boats, furniture, and more during the long winter months when farming would come to a halt. Growing up on a farm, I learned first-hand how vulnerable crops are to weather changes, insect invasions, and plant diseases.

I was inspired by my mother, who was a nurse, and this led me to study pre-med in college, with the intention of becoming either a doctor or a veterinarian. After two years of college, I was restless and wanted to see the larger world. So, I dropped out of school for a year and earned enough working on various farms to buy a round-trip plane ticket from Idaho to France. In 1963, I traveled to Paris to live as a student for a semester. After arriving, I learned that my residence was in the same student building as the chaplain—a Jesuit Priest by the name of Daniel Berrigan. Father Berrigan was a well-known anti-war, peace activist and, while living in Paris, we spoke countless times and three themes would always come up: the war in Viet Nam, racism in America and the world, and the importance of showing up in life fully and peacefully. Father Berrigan left a lasting impression on me—his deep commitment to peace and social justice, his active resistance to the war in Viet Nam, and the simple ways in which he lived.

After living in Europe for a half-year during a time of student social unrest, I realized that I was less motivated to become a traditional doctor. Instead of physical healing, I felt drawn to a life of social healing—but had no clear idea what forms that could take. After completing my undergraduate education, I began four years of graduate school at the University of Pennsylvania, where I earned an MBA from the Wharton School, and an MA in economic history.

After completing this graduate work, in 1972, I began my first white-collar job, working as a senior researcher on the staff of the "Presidential Commission on Population Growth and the American Future" in Washington, DC. It was an eye-opening experience for this farm boy to work on a presidential commission. Our mandate: to look ahead thirty years, from 1970 to 2000, and explore population growth and urbanization. Although the commission had a budget and lifespan for only two years, this was an invaluable introduction to research on the long-range future. It was also tremendous opportunity to observe politics at the White House level and to see how government works. I was surprised to see the extent to which policies are dominated by short-term considerations and the power of special interests.

Disillusioned, I left Washington for California to begin work as a senior social scientist in the "futures group" of the Stanford Research Institute (SRI International) think-tank. Over the next six years, I co-authored numerous studies of the long-range future: for example, *Anticipating Future National and Global Problems* (for the National Science Foundation), *Alternative Futures for Environmental Policy: 1975–2000* (for the Environmental Protection Agency), and *Limits to the Management of Large, Complex Systems* (for the President's Science Advisor). I also co-authored a pioneering study with Joseph Campbell and a small team of scholars titled *Changing Images of Man*. This research explored archetypes drawing humanity into a transforming future, and profoundly deepened my understanding of humanity's evolutionary journey. Taken together, these years of research made it clear we humans were on an unsustainable path and, within a few decades, would begin to so overconsume the Earth's resources that we would move into a condition of planetary breakdown and collapse. I saw how humanity would need to make deep changes if we were to avoid destroying the biosphere. At the same time, my inner growth was being catalyzed in surprising ways.

A remarkable opportunity emerged while working at SRI—to become a subject in psychic research that was just getting underway.

The U.S. government was beginning to fund its earliest research exploring humanity's intuitive skills and psychic potentials. The initial research began at SRI in the early 1970s, funded by NASA, and available to the public. I was fortunate to become one of four primary subjects and to participate in a wide range of experiments exploring both the "receiving" and "sending" aspects of consciousness. The receiving aspect included "remote viewing" or seeing places and people at a distance with direct intuition. The sending aspect included "psychokinesis" and involved intuitive engagement with physical systems. Over the course of three years, I learned a core lesson again and again: The world is alive and permeated with consciousness and subtle energy. Our physical body provides a stable foundation for learning about the nature of consciousness, which is not limited to our body, but extends into the universe as an ever-present intelligent knowing and aliveness. In turn, we are far bigger than our physical body and endowed with far more subtle capacities than I previously imagined. We are just beginning to use highly sensitive technologies to provide feedback and develop a "literacy of consciousness." Lessons learned in this laboratory work continue to inform my understanding a half-century later.

    I left SRI in 1977 and began focusing my efforts on becoming a "media activist." For decades, I had watched how the mass media dominate and orient the mass mind of entire civilizations. Our collective consciousness was being profoundly affected by both the huge amount of ads selling a materialistic mindset, and by the media ignoring key challenges such as climate change, poverty, and racism. I began doing non-partisan, community organizing in the San Francisco Bay Area with the goal of fostering mass media much more responsive to the needs of citizens. To accomplish this, we created a non-profit organization—Bay Voice—that challenged the licenses of the major broadcast TV stations in the San Francisco Bay Area on the grounds that they were not serving the legal rights of citizens to be informed. In 1987, Bay Voice collaborated with ABC-TV station to produce an historic, hour-long, prime-time "Electronic Town Meeting" viewed by more than 300,000 people, and included six votes from a scientific sample of citizens during the live TV program. The public gave the television station very strong, valuable feedback about their programming.

A contemporary expression of this work is the *Earth Voice* initiative described in this book, which will use internet technology, now accessible to a majority of citizens on the Earth, to create a planetary-scale voice for the Earth.

Writing and research have been a significant part of my work. For me, writing is much more than a mental exercise; it is a whole-body experience of feeling and digesting the meaning of something, so words then embody the felt experience that gives rise to them. Seeing and feeling how we are overconsuming the Earth, I began writing about simplicity of living in the mid 1970s. My book, **Voluntary Simplicity**: *Toward a Way of Life that is Outwardly Simple, Inwardly Rich,* was first published in 1981 and republished in 2009. My experience of working on the Changing Images of Man project felt incomplete and I invested nearly 15 years in writing my own version of this report—**Awakening Earth**: *Exploring the Evolution of Human Culture and Consciousness* was published in 1993. Seeing how slow we were in moving toward a more constructive and sustainable future, I wrote **Promise Ahead**: *A Vision of Hope and Action for Humanity's Future,* published in 2000. While involved in the parapsychology experiments in the early 1970s, I began writing about the nature of the universe as a living system permeated by consciousness and this culminated more than 30 years later in my book **The Living Universe**: *Where Are We? Who Are We? Where Are We Going?* published in 2009. In addition to these books, I have contributed chapters to more than two-dozen books, and published more than a hundred major articles. These decades of research and writing have all converged and contributed to the writing of **Choosing Earth**.

Throughout these decades, I have had the good fortune to travel to different parts of the world and give talks to diverse audiences on diverse themes. I have given more than 350 keynote speeches to different audiences ranging from business leaders, and non-profit organizations, to universities, film and media groups, religious organizations, and others. I have also been blessed to attend meetings and gatherings with people from all walks of life—including leaders, teachers, students, and workers.

In 2006, I was honored to receive Japan's "Goi Peace Award" in Tokyo in recognition of my contributions to a global "vision, consciousness, and lifestyle" that fosters a "more sustainable and spiritual culture." In 2001, I received an honorary Doctor of Philosophy from the California Institute of Integral Studies in recognition for my work for "ecological and spiritual transformation."

Looking back a half century, I can see how my professional career has led me to author this final book, *Choosing Earth*. My intention now is to bring this book and the companion documentary and courses into the world through collaborations, organizing, consulting, speaking, and teaching. Please visit these websites to learn more: my personal website: www.DuaneElgin.com and my professional website: www.ChoosingEarth.org

Endnotes

1. James Hillman, *Re-Visioning Psychology* (New York: Harper and Row, 1975), 16.
2. Robin Wall Kimmerer, *Braiding Sweetgrass* (Minneapolis, MN: Milkweed Editions, 2013), 359.
3. Alexis Pauline Gumbs, *Undrowned* (Chico, CA: AK Press, 2020), 15.
4. Mia Birdsong, *How We Show Up* (New York: Hachette Books, 2020), 38.
5. "The Beginning of the End," editors of the journal, *New Scientist*, October 13, 2018. https://www.newscientist.com/article/mg24031992-900-weve-missed-many-chances-to-curb-global-warming-this-may-be-our-last/
6. "The Report of The Commission on Population Growth and the American Future," https://www.population-security.org/rockefeller/001_population_growth_and_the_american_future.htm
7. Willis Harman and Peter Schwartz, *Assessment of Future National and International Problem Areas*, Prepared for the National Science Foundation, Contract NSF/STP76-02573, SRI Project 4676, February 1977. In addition to contributing to the overall report, I also authored an individual, 77-page report: *Limits to the Management of Large, Complex Systems*, published as a companion volume, February 1977.
8. Duane Elgin, ibid., a summary of this 77-page report on *Limits to the Management of Large, Complex Systems*. was published as the article: "Limits to Complexity: Are Bureaucracies Becoming Unmanageable," in *The Futurist,* December 1977. https://duaneelgin.com/wp-content/uploads/2014/11/Limits-to-Large-Complex-Systems.pdf
9. A summary description of this half-year of meditation in 1978, was included as an appendix in my book *Awakening Earth*. This book is available as a free download on my personal website: https://duaneelgin.com/wp-content/uploads/2016/03/AWAKENING-EARTH-e-book-2.0.pdf Insights from this meditation experience provided the foundation for exploring beyond the current materialist paradigm and these are described as a theory of "dimensional evolution." *Awakening Earth* presents the mid-2020s as the approximate time-frame for moving into the next, more spacious, dimensional context of a living universe paradigm and its view of reality, human identity, and evolutionary journey.
10. With gratitude to the Buddhist monk Thich Nat Hanh for offering this description.
11. Caroline Hickman, et. al., "Young people's voices on climate anxiety, government betrayal and moral injury: a global phenomenon." University of Bath, UK, September 14, 2021. https://papers.ssrn.com/sol3/papers.cfm?abstract_id=3918955

12 "Peoples' Climate Vote," UN Development Program (UNDP) and University of Oxford, January 2021, https://www.undp.org/publications/peoples-climate-vote#modal-publication-download.

13 "World Scientists' Warning to Humanity," *Union of Concerned Scientists*, 1992 onwards. https://www.ucsusa.org/resources/1992-world-scientists-warning-humanity

14 Ibid.

15 Owen Gaffney, "Quit Carbon, and Quick," *New Scientist*, January 5, 2019. https://www.sciencedirect.com/science/article/abs/pii/S0262407919300181

16 Eugene Linden, "How Scientists Got Climate Change So Wrong," *The New York Times*, November 8, 2019. https://www.nytimes.com/2019/11/08/opinion/sunday/science-climate-change.html Also:

"Climate Change Speed-Up," *Atmospheric Sciences & Global Change Research Highlights*, March 2015. Increasing temperature change over next several decades will accelerate, according to new research. Earth's temperature changes are happening faster than historical levels and are starting to speed up. https://www.pnnl.gov/science/highlights/highlight.asp?id=3931

"How fast is the climate changing? It's happened within one lifetime." David Wallace-Wells, climate journalist and author of *The Uninhabitable Earth*, explains: https://www.youtube.com/watch?v=RA4mIbQo52k

17 Although time-scales of events described as "abrupt" may vary dramatically, there is very disturbing evidence they can be on the timescale of years! For example: "Changes recorded in the climate of Greenland at the end of the Younger Dryas [roughly 11,800 years ago], as measured by ice-cores, imply a sudden warming of +10° C (+18° F) within a timescale of a few years." Grachev, A.M.; Severinghaus, J.P., Quaternary Science Reviews, March, 2005. "A revised +10±4° C magnitude of the abrupt change in Greenland temperature at the Younger Dryas termination using published GISP2 gas isotope data and air thermal diffusion constants." https://ui.adsabs.harvard.edu/abs/2005QSRv...24..513G/abstract

18 An exception is Sweden: Christian Ketels and K. Persson, "Sweden's ministry for the future: how governments should think strategically and act horizontally," *Centre for Public Impact*, November 29, 2018. https://www.centreforpublicimpact.org/swedens-ministry-for-the-future-how-governments-should-think-strategically-and-act-horizontally/

19 Gus Speth, quoted in the *Canadian Association of the Club of Rome*, March 27, 2016. https://canadiancor.com/scientists-dont-know/

20 John Vidal, "The Lost Decade: How We Awoke To Climate Change Only To Squander Every Chance To Act," *HuffPost*, December 30, 2019. https://www.huffpost.com/entry/lost-decade-climate-change-action-2020_n_5df7af92e4b0ae01a1e459d2

21 "Workers Flee and Thieves Loot Venezuela's Reeling Oil Giant," *The New York Times*, June 14, 2018. https://www.nytimes.com/2018/06/14/world/americas/venezuela-oil-economy.html

22 "*Gangs Rule Much of Haiti. For Many, It Means No Fuel, No Power, No Food,*" https://www.nytimes.com/2021/10/27/world/americas/haiti-gangs-fuel-shortage.html "Haiti descends into chaos, yet the world continues to look away," editorial board, Washington Post, November 21, 2021. https://www.washingtonpost.com/opinions/2021/10/31/haiti-descends-into-chaos-yet-world-continues-look-away/

23 See, for example: Future of Life Institute, https://futureoflife.org/background/existential-risk/

24 To illustrate, a "transhumanist" movement is underway in popular culture and has been described as "a social and philosophical movement devoted to promoting the research and development of robust human-enhancement technologies. Such technologies would augment or increase human sensory reception, emotive ability, or cognitive capacity, as well as radically improve human health and extend human life spans." https://en.wikipedia.org/wiki/Transhumanism

25 Although highly controversial, it is important to acknowledge the role of genetic editing for the future. Rewriting the code of life is swiftly becoming a technology that could rewrite humanity's evolutionary future — particularly within the time frame of a half century considered here. CRISPR is the gene editing tool that operates like the find-and-replace function in a word processor. Instead of requiring a massive science lab, this technology has become very easy to use and has fostered numerous, garage-scale gene entrepreneurs seeking to create and sell new gene lines to humanity. The World Health Organization has noted that *gene editing tools do not require exceptional biochemical knowledge or skills, nor significant funds, nor significant amounts of time.* Understandably, then, these tools have moved from large-scale sophisticated laboratories in universities to the garages and living rooms of "bio-hackers" who are working, virtually without regulation, to create new threads of life that are essentially impossible to undo. Genetic editing is a dual impact technology, which means it can bring both great benefit and great harm to the world.

The potential benefits of this technology are enormous. Genetic editing can help feed the world with disease-resistant and drought-tolerant plants. These tools can also be used to create designer humans with tolerance to high heat and stress, as well as resistance to many diseases. For example, it can go a long way toward curing roughly 7,000 human diseases that are caused by genetic mutations. It could make people more resistant to the AIDS virus and other diseases, such as sickle-cell anemia, cystic fibrosis, heart disease, leukemia, malaria resistance, and perhaps Alzheimer's. Another example of major benefit is responding to human sperm counts that have been dropping dramatically. If they fall to near zero, then the functional extinction of the human species becomes likely as we will be unable to reproduce ourselves. In turn, there would likely be concerted efforts to use genetic editing to produce more robust and resilient sperm that can survive the evolutionary pressures of our world in deep transition.

The harm that could come from this technology is also enormous. Apart from climate change, there are only two technologies that could rapidly kill billions of people: nuclear weapons and biological weapons. To illustrate, smallpox is one of the world's most contagious, disfiguring, and deadly diseases that has affected humans for thousands of years and has killed roughly 30 percent of those who get infected. Although smallpox has been eradicated from the Earth, scientists have discovered that it can be reassembled in a bio-hacker lab by putting together components available in the world. Genetic editing could also be used to engineer drug-resistant anthrax or highly transmissible influenza, and many more.

Gene editing is an evolutionary wildcard that could shift the direction of evolution in unknown directions. Historian and futurist Yuval Noah Harari writes in his controversial book *Homo Deus* that if we use this technology, humanity will begin to break the laws of natural selection that have shaped life for the past four-billion years and replace them with "laws of true intelligent design." Within a few decades, the Earth could be inhabited by genetically augmented humans whose great advantages could make them both essential and nearly unstoppable, thereby producing a bio-genetically stratified society. Each generation of "enhanced" humans could establish a new base line for upgrading the next generation, thereby producing radically different types of humans — but along which lines? If augmented capacities are grounded in the shallow paradigm of materialism, they seem likely to create a bleak future for humanity. To illustrate, Harari writes that genetically augmented humans will be honored for "the contribution they make to the data streams that various computer-assisted algorithms are using to generate value and create production."

The paradigm of materialism provides the foundation for this impoverished and shallow view of humanity's evolutionary potentials. Harari writes that, "In the future, we may see real gaps in physical and cognitive abilities opening between an upgraded upper class and the rest of society and that we could have "upgraded superhumans who dominate the world" thereby creating "a new superhuman caste that will abandon its liberal roots and treat normal humans no better than nineteenth-century Europeans treated Africans." In turn, he states the most ruthless evolutionary strategy might be to let go of the world's poor and unskilled, and dash forward with the augmented class only. Without a transcending ethical context for guiding this emerging bio-genetic revolution, there is an enormous danger of creating a new caste system — and a profoundly diminished and distorted future for humanity. (See Yuval Harari, *Homo Deus*, New York: Harper Collins, 2017, p. 352 - 355. Also: interview by Ezra Klein: "Yuval Harari, author of *Sapiens*," https://www.vox.com/2017/2/28/14745596/yuval-harari-sapiens-interview-meditation-ezra-klein.)

26 "We've missed many chances to curb global warming. This may be our last," editors of the journal, *New Scientist,* October 13, 2018. https://www.newscientist.com/article/mg24031992-900-weve-missed-many-chances-to-curb-global-warming-this-may-be-our-last/

27 Jared Diamond, *Collapse: How Societies Choose to Fail or Succeed,* New York: Penguin Group, 2005. Also: Diamond, "Easter's End," in *Discover Magazine,* December 31, 1995, https://www.discovermagazine.com/planet-earth/easters-end

28 Op. cit., *Collapse,* p. 109.

29 Ibid., p. 119.

30 Garry Kasparov and Thor Halvorssen, "Why the rise of authoritarianism is a global catastrophe," *Washington Post,* February 13, 2017. https://www.washingtonpost.com/news/democracy-post/wp/2017/02/13/why-the-rise-of-authoritarianism-is-a-global-catastrophe/

31 Maria Repnikova, "China's 'responsive' authoritarianism," *Washington Post,* November 27, 2019. https://www.washingtonpost.com/news/theworldpost/wp/2018/11/27/china-authoritarian/ Also: Paul Mozur and Aaron Krolik, "A Surveillance Net Blankets China's Cities, Giving Police Vast Powers," *The New York Times,* December 17, 2019. https://www.nytimes.com/2019/12/17/technology/china-surveillance.html?action=click&module=Top%20Stories&pgtype=Homepage

32 Nicholas Wright, "How Artificial Intelligence Will Reshape the Global Order," *Foreign Affairs,* July 10, 2018. https://www.foreignaffairs.com/articles/world/2018-07-10/how-artificial-intelligence-will-reshape-global-order?fa_anthology=1123571

33 Mathew Macwilliams, "Trump is an authoritarian. So are millions of Americans." *Politico,* September 23, 2020. https://www.politico.com/news/magazine/2020/09/23/trump-america-authoritarianism-420681 "Call it authoritarianism," *Vox,* Jun 15, 2021, https://www.vox.com/policy-and-politics/2021/6/15/22522504/republicans-authoritarianism-trump-competitive

34 Duane Elgin, *The Living Universe,* op. cit., 2009, p. 141-142.

35 "World Income Inequality Report," *World Inequality Lab,* December 2021. https://wid.world/news-article/world-inequality-report-2022/

36 Another indication of the danger ahead is described in the IPCC report on climate change and land. See: "The world has just over a decade to get climate change under control, U.N. scientists say." *Washington Post,* Chris Mooney and Brady Dennis, Oct. 7, 2018. There is no documented historic precedent for the scale of changes required, the body found. Here is an important response to the new IPCC report on "Climate Change and Land." 1.5° is the new 2°," Jennifer Morgan, executive director of Greenpeace International. Specifically, the document finds that instabilities in Antarctica and Greenland, which could usher in sea-level rise measured in feet rather

than inches, "could be triggered around 1.5°C to 2°C of global warming." Moreover, the total loss of tropical coral reefs is at stake because 70 percent to 90 percent are expected to vanish at 1.5° Celsius, the report finds. At 2°C, that number grows to more than 99 percent. The report clearly documents that a warming of 1.5°C would be very damaging and that 2°C—which used to be considered a reasonable goal—could produce intolerable consequences in parts of the world. https://www.ipcc.ch/report/srccl/ Also:

Updated IPCC report: "New U.N. climate report: Massive change already here for world's oceans and frozen regions," Chris Mooney and Brady Dennis, *Washington Post*, September 25, 2019. https://www.washingtonpost.com/climate-environment/2019/09/25/new-un-climate-report-massive-change-already-here-worlds-oceans-frozen-regions/

Special Report on the Ocean and Cryosphere in a Changing Climate, Intergovernmental Panel on Climate Change, https://www.ipcc.ch/srocc/ To download, go to: https://www.ipcc.ch/srocc/download-report/

37 An example of damage from sea-level rise is the severe erosion of the world's beaches: Half of the world's beaches could disappear by the end of the century, and by 2050 some coastlines could be unrecognizable from what we see today. Michalis I. Vousdoukas, et. al., "Sandy coastlines under threat of erosion," *Nature: Climate Change*, March 2, 2020. https://www.nature.com/articles/s41558-020-0697-0

38 IPCC Special Report on the Ocean and Cryosphere in a Changing Climate, IPCC, September 25, 2019. https://www.ipcc.ch/srocc/ To download, go to: https://www.ipcc.ch/srocc/download-report/ https://climatenexus.org/climate-change-news/ipcc-oceans-ice-systems-climate-impacts/

39 "Sea levels set to keep rising for centuries even if emissions targets met," *The Guardian,* November 6, 2019. The lag time between rising global temperatures and the impact of coastal inundation means that the world will be dealing with ever-rising sea levels into the 2300s, regardless of prompt action to address the climate crisis, according to the new study. https://www.theguardian.com/environment/2019/nov/06/sea-level-rise-centuries-climate-crisis See the study "Attributing long-term sea-level rise to Paris Agreement emission pledges,": https://www.pnas.org/content/early/2019/10/31/1907461116 Also: Zeke Hausfather, "Common Climate Misconceptions: Atmospheric Carbon Dioxide," *Yale Climate Connections*, December 16, 2010. This study found that while a good portion of greenhouse gas emissions could be removed from the atmosphere in a few decades, even if emissions were somehow ceased immediately, about 10 percent would continue warming Earth for thousands of years. This 10 percent is significant, because even a small increase in atmospheric greenhouse gases can have a large impact on ice sheets and sea level if it persists over the millennia. Even more important: The biggest danger is not global warming, it is the extreme weather produced by moving past tipping points which, in turn, lead to catastrophic famine and immense civic unrest. https://www.yaleclimateconnections.org/2010/12/common-climate-misconceptions-atmospheric-carbon-dioxide/

40 "BP Statistical Review of World Energy," *British Petroleum*, (68th edition), 2019. https://www.bp.com/content/dam/bp/business-sites/en/global/corporate/pdfs/energy-economics/statistical-review/bp-stats-review-2019-full-report.pdf

41 "Hothouse Earth Fears," *New Scientist*, August 11, 2018. https://www.sciencedirect.com/journal/new-scientist/vol/239/issue/3190 "For most of the past half billion years, Earth has been much hotter than today, with no permanent ice at the poles: the hothouse Earth state. Then, roughly three million years ago, as CO_2 levels fell, temperatures began oscillating between two cooler states: ice ages with great ice sheets covering much land in the northern hemisphere and interglacial periods like the present. With CO_2 increases we might be on the brink of pushing the planet out of the present interglacial state and into the hothouse Earth state. The consequences are beyond catastrophic." Also:

McGrath, "Climate change: 'Hothouse Earth' risks even if CO_2 emissions slashed," *BBC*, August 5, 2018. https://www.bbc.com/news/science-environment-45084144

"New Climate Risk Classification Created to Account for Potential 'Existential' Threats," *Scripps Institute of Oceanography*, September 14, 2017. "A temperature increase greater than 3°C (5.4°F) could lead to what the researchers term "catastrophic" effects, and an increase greater than 5°C (9°F) could lead to "unknown" consequences, which they describe as beyond catastrophic, including potentially existential threats. The specter of existential threats is raised to reflect the grave risks to human health and species extinction from warming beyond 5°C, which has not been experienced for at least the past 20 million years." https://scripps.ucsd.edu/news/new-climate-risk-classification-created-account-potential-existential-threats

Will Steffen, et. al., "Trajectories of the Earth System in the Anthropocene," *PNAS: Proceedings of the National Academy of Sciences*, August 14, 2018. "We explore the risk that self-reinforcing feedbacks could push the Earth system toward a planetary threshold that, if crossed, could prevent stabilization of the climate at intermediate temperature rises and cause continued warming on a 'Hothouse Earth' pathway, even as human emissions are reduced. Crossing the threshold would lead to a much higher global average temperature than any interglacial in the past 1.2 million years and to sea levels significantly higher than at any time in the Holocene." https://doi.org/10.1073/pnas.1810141115

42 "Climate Change: How Do We Know?" *NASA: Global Climate Change, Vital Signs of the Planet,* 2019. See the evidence here: https://climate.nasa.gov/evidence/ See scientific consensus regarding climate warming here: https://climate.nasa.gov/scientific-consensus/ Also:

"Climate change: Disruption, risk and opportunity," *Science Direct* (originally published in *Global Transitions*, Volume 1, 2019, pp. 44-49). The study concludes: Climate change is disruptive because humans have adapted to a narrow range of environmental conditions. Change is particularly risky in the presence of low predictability, large-scale, rapid onset and lack of reversibility. https://doi.org/10.1016/j.glt.2019.02.001

"Global Warming Science: The science is clear. Global warming is happening." *Union of Concerned Scientists*, 2019. https://www.ucsusa.org/our-work/global-warming/science-and-impacts/global-warming-science

Op. cit., *IPCC Special Report on* Oceans *and the Cryosphere*, September 25, 2019.

Bob Berwyn, "Ocean Warming Is Speeding Up, with Devastating Consequences, Study Shows," *Inside Climate News*. January 14, 2020. In 25 years, the oceans have absorbed heat equivalent to the energy of 3.6 billion Hiroshima-size atom bomb explosions, the study's lead author said. https://insideclimatenews.org/news/14012020/ocean-heat-2019-warmest-year-argo-hurricanes-corals-marine-animals-heatwaves

Sabrina Shankman, "Dead Birds Washing Up by the Thousands Send a Warning About Climate Change," *Inside Climate News*, January 15, 2020. A new study unravels the mystery of what caused so many of these normally resilient seabirds to starve amid an ocean heat wave fueled in part by global warming. https://insideclimatenews.org/news/15012020/seabird-death-ocean-heat-wave-blob-pacific-alaska-common-murre

43 "Urgent health challenges for the next decade," *WHO (World Health Organization)*, January 13, 2020. https://www.who.int/news-room/photo-story/photo-story-detail/urgent-health-challenges-for-the-next-decade

44 "Powerful actor, high impact bio-threats — initial report," *Wilton Park/UK*, November 9, 2018. https://www.wiltonpark.org.uk/wp-content/uploads/WP1625-Summary-report.pdf Also:

Nafeez Ahmed, "Coronavirus, Synchronous Failure and the Global Phase-Shift," *Insurge Intelligence*, March 2, 2020. https://medium.com/insurge-intelligence/coronavirus-synchronous-failure-and-the-global-phase-shift-3f00d4552940

Jennifer Zhang, "Coronavirus Response Shows the World May Not Be Ready for Climate-Induced Pandemics," *Columbia University*, February 24, 2020. https://blogs.ei.columbia.edu/2020/02/24/coronavirus-climate-induced-pandemics/

Brian Deese and Ronald Klain, "Another deadly consequence of climate change: The spread of dangerous diseases," *Washington Post,* May 30, 2017. https://www.washingtonpost.com/opinions/another-deadly-consequence-of-climate-change-the-spread-of-dangerous-diseases/2017/05/30/fd3b8504-34b1-11e7-b4ee-434b6d506b37_story.html

I appreciate the insights of Sandy Wiggins in differentiating between the challenges of responding to pandemics and those of climate change.

45 Another study concludes that already: "Two-thirds of the global population (4.0 billion people) live under conditions of severe water scarcity at least 1 month of the year." https://www.seametrics.com/blog/future-water/ Also:

Mesfin M. Mekonnen and Arjen Y. Hoekstra, "Four billion people facing severe water scarcity," Science Advances, February 12, 2016. https://advances.sciencemag.org/content/2/2/e1500323.full

Another study found that between 1995 and 2025 the areas affected by "severe water stress" expand and intensify, and the number of people living in these areas also grows from 2.1 to 4.0 billion people. They state: "continuing stress on water resources increases the risk that simultaneous water shortages might occur around the world and even trigger a kind of global water crisis." "World Water in 2025: Global modeling and scenario analysis for the World Commission on Water for the coming century," Joseph Alcamo, Thomas Henrichs, Thomas Rösch, *Center for Environmental Systems Research University* of Kassel, February 2000. file:///Users/duaneelgin/Downloads/World%20Water%20in%202025%20(2).pdf

46 "The Water Crisis," Water.org, 2019. https://water.org/our-impact/water-crisis/

47 "World Water Development Report," 2019. https://www.unwater.org/publications/world-water-development-report-2019/ Also: https://water.org/our-impact/water-crisis/

48 The number of undernourished people in the world has been on the rise since 2015 and is back to levels seen in 2010–2011. http://www.fao.org/state-of-food-security-nutrition/en/ Also:

"The Hungry Planet: Global Food Scarcity in the 21st Century," Wilson Center Staff, August 16, 2011. https://www.newsecuritybeat.org/2011/08/the-hungry-planet-global-food-scarcity-in-the-21st-century/

Julian Cribb, "The coming famine: risks and solutions for global food security," October 21, 2009. https://www.dpi.nsw.gov.au/__data/assets/pdf_file/0013/304402/Julian-Cribb---Global-Food-Oct09.pdf

49 Nafeez Ahmed, "West's 'Dust Bowl' Future now 'Locked In, as World Risks Imminent Food Crisis," *Insurge Intelligence*, January 6, 2020. https://www.resilience.org/stories/2020-01-06/wests-dust-bowl-future-now-locked-in-as-world-risks-imminent-food-crisis/

50 Anup Shah, "Poverty Facts and Stats," *Global Issues*, Updated January 7, 2013. http://www.globalissues.org/article/26/poverty-facts-and-stats#src1 Also:

Anup Shah, "Poverty Around The World," *Global Issues,* November 12, 2011. http://www.globalissues.org/print/article/4#WorldBanksPovertyEstimatesRevised

51 Julian Cribb, "The coming famine: risks and solutions for global food security," October 21, 2009. https://www.dpi.nsw.gov.au/__data/assets/pdf_file/0013/304402/Julian-Cribb---Global-Food-Oct09.pdf

52 "Our Food Systems Are in Crisis," *Scientific American*, October 15, 2019. https://blogs.scientificamerican.com/observations/our-food-systems-are-in-crisis/

53 Izabella Koziell, "Migration, Agriculture and Climate Change," *Food and Agricultural Organization of the United Nations*, 2017. http://www.fao.org/3/I8297EN/i8297en.pdf

54 See report "Nature's Dangerous Decline 'Unprecedented'; Species Extinction Rates 'Accelerating'" *Intergovernmental Science-Policy Platform on Biodiversity and Ecosystem Services (IPBES),* May 22, 2019. https://www.ipbes.net/news/Media-Release-Global-Assessment Also: https://www.washingtonpost.com/climate-environment/2019/05/06/one-million-species-face-extinction-un-panel-says-humans-will-suffer-result/

55 "Plummeting insect numbers 'threaten collapse of nature,'" in *The Guardian*, February 10, 2019. https://www.theguardian.com/environment/2019/feb/10/plummeting-insect-numbers-threaten-collapse-of-nature A growing number of studies are sounding the alarm that **insects** around the world are in crisis. For example, one study in Germany found a 76 percent decrease in flying insects in just the past few decades. Another study of rainforests in Puerto Rico found insects had declined as much as 60-fold. Also:

Damian Carrington, "Car 'splatometer' tests reveal huge decline in number of insects," *The Guardian*, February 12, 2020. Research shows insect populations at sites in Europe has plunged by up to 80 percent in two decades. https://www.theguardian.com/environment/2020/feb/12/car-splatometer-tests-reveal-huge-decline-number-insects

Damian Carrington, "Insect apocalypse' poses risk to all life on Earth, conservationists warn," *The Guardian*, November 13, 2019. Report claims 400,000 insect species face extinction amid heavy use of pesticides. https://www.theguardian.com/environment/2019/nov/13/insect-apocalypse-poses-risk-to-all-life-on-earth-conservationists-warn

Dave Goulson, "Insect declines and why they matter," commissioned by the *South West Wildlife Trusts*, 2019. ". . . recent evidence suggests that abundance of insects may have fallen by 50 percent or more since 1970. This is troubling, because insects are vitally important, as food, pollinators and recyclers amongst other things." https://www.somersetwildlife.org/sites/default/files/2019-11/FULL%20AFI%20REPORT%20WEB1_1.pdf https://doi.org/10.1016/j.biocon.2019.01.020

56 "Pollinators Help One-third Of The World's Food Crop Production," *Science Daily*, October 26, 2009. https://www.sciencedaily.com/releases/2006/10/061025165904.htm Bees are the primary initiators of reproduction among plants, as they transfer pollen from the male stamens to the female pistils. Also:

57 Carl Zimmer, "Birds Are Vanishing from North America," *The New York Times*, September 19, 2019. https://www.nytimes.com/2019/09/19/science/bird-populations-america-canada.html

58 Kenneth Rosenberg, et. al., "Decline of the North American avifauna," *Science*, October 4, 2019. https://science.sciencemag.org/content/366/6461/120

59 J. Emmett Duffy, et. al., "Science study predicts collapse of all seafood fisheries by 2050," *Stanford Report*, November 2, 2006. https://news.stanford.edu/news/2006/november8/ocean-110806.html "All species of wild seafood will collapse within 50 years, according to a new study by an international team of ecologists and economists. . . . Based on current global trends, the authors predicted that every species of wild-caught seafood—from tuna to sardines—will collapse by the year 2050. 'Collapse' was defined as a 90 percent depletion of the species' baseline abundance." Also:

Jeff Colarossi, "Climate Change And Overfishing Are Driving The World's Oceans To The 'Brink Of Collapse,'" *Think Progress*, 2015. https://thinkprogress.org/climate-change-and-overfishing-are-driving-the-worlds-oceans-to-the-brink-of-collapse-2d095e127640/ "Within a single generation, human activity has severely damaged almost every aspect of our global oceans. That's the finding of a new *World Wildlife Fund* study, which revealed that marine populations have declined 49 percent between 1970 and 2012. . . . The picture is now clearer than ever: humanity is collectively mismanaging the ocean to the brink of collapse."

"Living Blue Planet Report: Species, habitats and human well-being," *World Wildlife Fund*, 2015. http://assets.wwf.org.uk/downloads/living_blue_planet_report_2015.pdf?_ga=1.259860873.2024073479.1442408269

Ivan Nagelkerken and Sean D. Connell, "Global alteration of ocean ecosystem functioning due to increasing human CO_2 emissions," *PNAS: Proceedings of the National Academy of Sciences*, October 27, 2015. https://doi.org/10.1073/pnas.1510856112

60 Adam Vaughan, "Humanity driving 'unprecedented' marine extinction," *The Guardian*, September 14, 2016. https://www.theguardian.com/environment/2016/sep/14/humanity-driving-unprecedented-marine-extinction The study can be found here: "Ecological selectivity of the emerging mass extinction in the oceans," *Science*, September 16, 2016. https://science.sciencemag.org/content/353/6305/1284

61 "Saving Life on Earth: a plan to halt the global extinction crisis," *Center for Biological Diversity*, January 2020. https://www.biologicaldiversity.org/programs/biodiversity/elements_of_biodiversity/extinction_crisis/pdfs/Saving-Life-On-Earth.pdf

62 Current U.N. World Population estimates. https://www.worldometers.info/world-population/

63 Rob Smith, "These will be the world's most populated countries by 2100," *World Economic Forum*, February 29, 2018. https://www.weforum.org/agenda/2018/02/these-will-be-the-worlds-most-populated-countries-by-2100/Also: Jeff Desjardins, "The world's biggest countries, as you've never seen them before," *World Economic Forum*, October 4, 2017. https://

www.weforum.org/agenda/2017/10/the-worlds-biggest-countries-as-youve-never-seen-them-before

64 World Population Growth. Sources: *Population Division of the Department of Economic and Social Affairs of the United Nations Secretariat*, 2013 and World Population Prospects The 2012 Revision, New York, *United Nations*. Less-developed regions: Africa, Asia (excluding Japan), Latin American and the Caribbean, and Oceana (excluding Australian and New Zealand). More-developed regions: Europe, North America (Canada and the United States), Japan, Australia, and New Zealand. https://kids.britannica.com/students/assembly/view/171828

65 Bradshaw and Barry Brook, "A killer plague wouldn't save the planet from us," *New Scientist,* November 1, 2014. An approximate carrying capacity of the Earth can be found in the article. The authors estimate a sustainable human population, given current Western consumption patterns and technologies, would be between 1 and 2 billion people. Also:

Another perspective on the Earth's carrying capacity is offered by Christopher Tucker, *A Planet of 3 Billion*, Atlas Observatory Press, August, 2019. http://planet3billion.com/index.html

Visionary scientist, James Lovelock, believes the Earth's population will fall to as few as 500 million by 2100, with most of the survivors living in the far northern latitudes—Canada, Iceland, Scandinavia, the Arctic Basin. See the interview: Jeff Goodell, "Hothouse Earth Is Merely the Beginning of the End," *Rolling Stone* magazine, August 9, 2018. https://www.rollingstone.com/politics/politics-features/hothouse-earth-climate-change-709470/

4 Degrees Hotter, A Climate Action Centre Primer, February 2011. Melbourne, Australia. https://www.climatecodered.org/2011/02/4-degrees-hotter-adaptation-trap.html The study quotes Professor Kevin Anderson, director of the *Tyndall Centre for Climate Change*, who "believes only around 10 percent of the planet's population—around half a billion people—will survive if global temperatures rise by 4°C. He said the consequences were "terrifying." "For humanity, it's a matter of life or death," he said. "We will not make all human beings extinct, as a few people with the right sort of resources may put themselves in the right parts of the world and survive. But I think it's extremely unlikely that we wouldn't have mass death at 4°C." In 2009 Professor Hans Joachim Schellbhuber, director of the *Potsdam Institute*, and one of Europe's most eminent climate scientists, told his audience that at 4°C, population ". . . carrying capacity estimates (are) below 1 billion people."

"Carrying capacity," *Wikipedia*, 2019. "Several estimates of the carrying capacity have been made with a wide range of population numbers. A 2001 U.N. report said that two-thirds of the estimates fall in the range of 4 billion to 16 billion with unspecified standard errors, with a median of about 10 billion. More recent estimates are much lower, particularly if non-renewable resource depletion and increased consumption are considered." https://en.wikipedia.org/wiki/Carrying_capacity

"How many people can Earth actually support?" *Australian Academy of Science*, 2019. https://www.science.org.au/curious/earth-environment/how-many-people-can-earth-actually-support "If everyone on Earth lived like a middle class American, then the planet might have a carrying capacity of around 2 billion." However, if people consumed only what they actually needed, then the Earth could potentially support a much higher figure.

Marian Starkey, "What is the Carrying Capacity of Earth?" *Population Connection*, April 13, 2017. https://populationconnection.org/blog/carrying-capacity-earth/ "Already, we're consuming the Earth's renewable resources at one and a half times the sustainable rate. And that's with billions of people living in poverty, consuming next to nothing. Imagine what would happen if desperately poor people were fortunate enough to live a middleclass lifestyle. And then imagine what would happen if poor people joined the middle class, *and* the human population grew from today's 7.5 billion to 9, 10, or 11 billion."

Andrew D. Hwang, "The human population is 7.5 billion and counting—a mathematician counts how many humans the Earth can actually support," *Business Insider*, July 10, 2018. https://www.businessinsider.com/how-many-people-earth-can-hold-before-runs-out-resources-2018-7 According to the *Worldwatch Institute*, an environmental think-tank, the Earth has 1.9 hectares of land per person for growing food and textiles for clothing, supplying wood and absorbing waste. The average American uses about 9.7 hectares. These data alone suggest the Earth can support at most one-fifth of the present population, 1.5 billion people, at a U.S. standard of living. The Earth supports industrialized standards of living only because we are drawing down the "savings account" of non-renewable resources, including fertile topsoil, drinkable water, forests, fisheries, and petroleum.

Natalie Wolchover, "How Many People Can Earth Support?," *Live Science*, October 11, 2011. https://www.livescience.com/16493-people-planet-earth-support.html "10 billion people is the uppermost population limit where food is concerned. Because it's extremely unlikely that everyone will agree to stop eating meat, E.O. Wilson thinks the maximum carrying capacity of the Earth based on food resources will most likely fall short of 10 billion."

"It is not the number of people on the planet that is the issue—but the number of consumers and the scale and nature of their consumption," says David Satterthwaite, a senior fellow at the International Institute for Environment and Development in London. He quotes Gandhi: "The world has enough for everyone's need, but not enough for everyone's greed." . . . The real concern would be if the people living in these areas decided to demand the lifestyles and consumption rates currently considered normal in high-income nations; something many would argue is only fair. . . . Only when wealthier groups are prepared to adopt low-carbon lifestyles, and to permit their governments to support such a seemingly unpopular move, will we reduce the pressure on global climate, resource and waste issues. . . . For the foreseeable future, Earth is our only home and we must find a way to live on it sustainably. It seems clear that that requires scaling back our consumption, in particular a transition to low-carbon lifestyles, and improving the status of women worldwide. Only when we have done these things will we really be able to estimate how many people our planet can sustainably hold."

"One Planet, How Many People? A Review of Earth's Carrying Capacity," *UNEP*, June 2102. https://na.unep.net/geas/archive/pdfs/geas_jun_12_carrying_capacity.pdf While there is a wide range to the estimates of Earth's carrying capacity, the greatest concentration of estimates falls between 8 and 16 billion people (3). Global population is fast approaching the low end of that range and is expected to get well into it at around 10 billion by the end of the century.

66 Ecological Footprint, https://www.footprintnetwork.org/our-work/ecological-footprint/

67 "Consumer Spending Trends and Current Statistics," Kimberly Amadeo, *The Balance*, June 27, 2019. https://www.thebalance.com/consumer-spending-trends-and-current-statistics-3305916 Also:

"Consumer Spending and the Economy," Hale Stewart, *The New York Times*, September 9, 2010. "The U.S. economy is predominantly driven by consumer spending, which accounts for approximately 70 percent of all economic growth. But if consumers are to continue to drive the economy, they must be in a sound financial position; if they become overburdened with debt, they are not able to maintain their position as the primary driver of economic growth." https://fivethirtyeight.blogs.nytimes.com/2010/09/19/consumer-spending-and-the-economy/

68 "Climate change: Big lifestyle changes 'needed to cut emissions'," Roger Harrabin, *BBC*, August 2019. https://www.bbc.com/news/science-environment-49499521

69 Report has been compiled by the World Meteorological Organization under the auspices of the Science Advisory Group of the U.N. Climate Action Summit 2019. https://wedocs.unep.org/bitstream/handle/20.500.11822/30023/climsci.pdf?sequence=1&isAllowed=y

70 Katherine Rooney, "Climate change will shrink these economies fastest," *World Economic Forum*, September 30, 2019. https://www.weforum.org/agenda/2019/09/climate-change-shrink-these-economies-fastest/

71 Nicholas Stern, "Climate change will force us to redefine economic growth," *World Economic Forum*, July 11, 2018. https://www.weforum.org/agenda/2018/07/here-are-the-economic-reasons-to-act-on-climate-change-immediately

72 Paul Buchheit, "These 6 Men Have as Much Wealth as Half the World's Population," *Common Dreams*, February 20, 2017. https://www.ecowatch.com/richest-men-in-the-world-2274065153.html

73 "Oxfam says wealth of richest 1% equal to other 99%." *BBC*, January 16, 2016. https://www.bbc.com/news/business-35339475

74 David Leonhardt, "The Rich Really Do Pay Lower Taxes Than You," *The New York Times*, October 6, 2019. https://www.nytimes.com/interactive/2019/10/06/opinion/income-tax-rate-wealthy.html?action=click&module=Opinion&pgtype=Homepage

75 Jason Kickel, "Global inequality may be much worse than we think," *The Guardian*, April 8, 2016. "Global inequality is worse than at any time since the 19th century. . . . It doesn't matter how you slice it; global inequality is getting worse. Much worse. Convergence theory turned out to be wildly incorrect. Inequality doesn't disappear automatically; it all depends on the balance of political power in the global economy. As long as a few rich countries have the power to set the rules to their own advantage, inequality will continue to worsen." https://www.theguardian.com/global-development-professionals-network/2016/apr/08/global-inequality-may-be-much-worse-than-we-think

76 Isabel Ortiz, "Global Inequality: Beyond the Bottom Billion," *UNICEF*, Working Paper, April, 2011. https://childimpact.unicef-irc.org/documents/view/id/120/lang/120_Global_Inequality_REVISED_-_5_July.pdf See Figure 7 for the "champagne glass" representation of inequities derived from the *United Nations Human Development Report* published in 1992 and published in Oxford University Press, 1992. Another version of the widely used "champagne glass" representation of inequalities is shown as Figure 1 in the report; "Extreme Carbon Inequality: why the Paris climate deal must put the poorest, lowest emitting and most vulnerable people first," *Oxfam Media Briefing, Oxfam.org*, December 2, 2015. https://oi-files-d8-prod.s3.eu-west-2.amazonaws.com/s3fs-public/file_attachments/mb-extreme-carbon-inequality-021215-en.pdf?te=1&nl=climate-fwd:&emc=edit_clim_20191113?campaign_id=54&instance_id=13827&segment_id=18753&user_id=d0fffc2fcb270a87206ab8a9cc08a01f&regi_id=63360062

77 "Extreme Carbon Inequality," ibid.

78 "Climate Justice," *Wikipedia*, https://en.wikipedia.org/wiki/Climate_justice

79 Andrew Hoerner and Nia Robinson, "A Climate of Change: African Americans, Global Warming, and a Just Climate Policy for the US," *Environmental Justice & Climate Change Initiative*, 2008. https://www.reimaginerpe.org/cj/hoerner-robinson

80 Moira Fagan, et. al., "A look at how people around the world view climate change," *PEW Research*, April 18, 2019. https://www.pewresearch.org/fact-tank/2019/04/18/a-look-at-how-people-around-the-world-view-climate-change/

81 Ibid., 2019.

82 I recognize how this terminology can be problematic because it assumes that the direction currently "developed" nations have taken (toward overconsumption and hyper-individualization) is the agreed-upon goal,

and that "developing" nations are simply lagging in their achievement of that goal.

83 "Scientific Consensus: Earth's Climate is Warming," *NASA: Global Climate Change, Vital Signs of the Planet,* 2019. See the evidence here: https://climate.nasa.gov/evidence/ See scientific consensus regarding climate warming here: https://climate.nasa.gov/scientific-consensus/ Also:

"Climate change: Disruption, risk and opportunity," *Science Direct* (originally published in *Global Transitions*, Volume 1, 2019). The study concludes: Climate change is disruptive because humans have adapted to a narrow range of environmental conditions. Change is particularly risky in the presence of low predictability, large-scale rapid onset, and lack of reversibility. https://doi.org/10.1016/j.glt.2019.02.001

"Global Warming Science: The science is clear. Global warming is happening." *Union of Concerned Scientists*, 2019. https://www.ucsusa.org/our-work/global-warming/science-and-impacts/global-warming-science

84 Timothy M. Lenton, et. al., "Climate tipping point—too risky to bet against," *Nature*, November 27, 2019. https://www.nature.com/articles/d41586-019-03595-0 Also:

Arthur Neslen, "By 2030, We Will Pass the Point Where We Can Stop Runaway Climate Change," HuffPost, September 5, 2018, https://www.huffingtonpost.com/entry/runaway-climate-change-2030-report_us_5b8ecba3e4b0162f4727a09f

The 2030s may be a period of high instability in climate trends—perhaps involving a climatological "whiplash." For example, a 2015 study predicted cooling rather than warming in this decade: "Solar activity predicted to fall 60 percent in 2030s, to mini-ice age levels: Sun driven by double dynamo," July 9, 2015, *Royal Astronomical Society*, reported in *Science Daily*. https://www.sciencedaily.com/releases/2015/07/150709092955.htm

Alexander Robinson, et al., "Multistability and critical thresholds of the Greenland ice sheet," Nature Climate Change March 1, 2012. "... the Greenland ice sheet is more sensitive to long-term climate change than previously thought. We estimate that the warming threshold leading to an essentially ice-free state is in the range of 0.8–3.2°C, with a best estimate of 1.6°C." https://www.nature.com/articles/nclimate1449#citeas

Michael Marshall, "Major methane release is almost inevitable," *New Scientist*, February 21, 2013. "We are on the cusp of a tipping point in the climate. If the global climate warms another few tenths of a degree, a large expanse of the Siberian permafrost will start to melt uncontrollably." https://www.newscientist.com/article/dn23205-major-methane-release-is-almost-inevitable/#ixzz5zQ199XTi

"Jessica Corbett, "'Boiling with methane': Scientists reveal 'truly terrifying' sign of climate change under the Arctic Ocean," *Common Dreams,* October 9, 2019. https://www.alternet.org/2019/10/boiling-with-methane-scientists-reveal-truly-terrifying-sign-of-climate-change-under-the-arctic-ocean/

85 "Temperature rise is 'locked-in' for the coming decades in the Arctic," *UNEP*, March 12, 2019. "Even if existing Paris Agreement commitments are met, winter temperatures over the Arctic Ocean will increase 3-5°C by mid-century compared to 1986-2005 levels. Thawing permafrost could wake 'sleeping giant' of more greenhouse gases, potentially derailing global climate goals." https://www.unenvironment.org/news-and-stories/press-release/temperature-rise-locked-coming-decades-arctic Also:

Steffen, et. al., "Trajectories of the Earth System in the Anthropocene," *PNAS*, July 6, 2018. This study explores: Hothouse Earth and how runaway global warming threatens habitability of the planet for humans. https://www.pnas.org/content/115/33/8252

86 "An unexpected surge in global atmospheric methane is threatening to erase the anticipated gains of the Paris Climate Agreement. Previously stable global methane levels have unexpectedly surged in recent years. See: Benjamin Hmiel, et.al., "Preindustrial 14CH_4 indicates greater anthropogenic fossil CH_4 emissions," *Nature*, February 19, 2020. https://www.nature.com/articles/s41586-020-1991-8 This study shows that scientists and governments have been greatly underestimating emissions of the powerful greenhouse gas methane from oil and gas operations. Also:

Nisbet et al. "Very Strong Atmospheric Methane Growth in the 4 Years 2014–2017: Implications for the Paris Agreement," *Global Biogeochemical Cycles*. March 2019. https://doi.org/10.1029/2018GB006009 See the summary article in *Climate Nexus* here: https://climatenexus.org/climate-change-news/methane-surge/

87 Hubau Wannes, et al., "Asynchronous carbon sink saturation in African and Amazonian tropical forests," *Nature*, March 5, 2020. https://www.nature.com/articles/s41586-020-2035-0 Also:

Fiona Harvey, "Tropical forests losing their ability to absorb carbon, study finds," *The Guardian*, March 4, 2020. https://www.theguardian.com/environment/2020/mar/04/tropical-forests-losing-their-ability-to-absorb-carbon-study-finds

88 Stewart Patrick, "The Coming Global Water Crisis," *The Atlantic*, May 9, 2012. https://www.theatlantic.com/international/archive/2012/05/the-coming-global-water-crisis/256896/ Also:

William Wheeler, "Global water crisis: too little, too much, or lack of a plan?," *Christian Science Monitor*, December 2, 2012. https://www.csmonitor.com/World/Global-Issues/2012/1202/Global-water-crisis-too-little-too-much-or-lack-of-a-plan

89 Gilbert Houngbo, "The United Nations world water development report 2018: nature-based solutions for water," *UNESCO*, 2018. https://unesdoc.unesco.org/ark:/48223/pf0000261424

90 Stephen Leahy, "From Not Enough to Too Much, the World's Water Crisis Explained," *National Geographic*, March 22, 2018. https://www.

nationalgeographic.com/news/2018/03/world-water-day-water-crisis-explained/

91 Paul Salopek, "Historic water crisis threatens 600 million people in India," *National Geographic*, October 19, 2018. https://www.nationalgeographic.com/culture/water-crisis-india-out-of-eden/?cmpid=org=ngp::mc=crm-email::src=ngp::cmp=editorial::add=Science_20200129&rid=51139F7FFEE4083137CDD6D1FF5C57FF

92 Dan Charles, "5 Major Crops In The Crosshairs Of Climate Change," *NPR*, October 25, 2018. https://www.npr.org/sections/thesalt/2018/10/25/658588158/5-major-crops-in-the-crosshairs-of-climate-change Also:

Sean Illing, "The climate crisis and the end of the golden era of food choice," *Vox*, June 24, 2019. https://www.vox.com/the-highlight/2019/6/17/18634198/food-diet-climate-change-amanda-little

Rachel Nuwer, "Here's how climate change will affect what you eat," *BBC*, December 28, 2015. https://www.bbc.com/future/article/20151228-heres-how-climate-change-will-affect-what-you-eat

Nicholas Thompson, "The Most Delicious Foods Will Fall Victim to Climate Change," *Wired*, June 13, 2019. https://www.wired.com/story/the-most-delicious-foods-will-fall-victim-to-climate-change/

Ian Burke, "29 of Your Favorite Foods That Are Threatened by Climate Change," *Saveur*, June 7, 2017. https://www.saveur.com/climate-change-ingredients/

Daisy Simmons, "A brief guide to the impacts of climate change on food production," *Yale Climate Connections*, September 18, 2019. https://www.yaleclimateconnections.org/2019/09/a-brief-guide-to-the-impacts-of-climate-change-on-food-production/

Ilima Loomis "Get ready to eat differently in a warmer world," *Science News for Students*, May 23, 2019. https://www.sciencenewsforstudents.org/article/climate-change-global-warming-food-eating

Peter Schwartzstein, "Indigenous farming practices failing as climate change disrupts seasons," *National Geographic*, October 14, 2019. https://www.nationalgeographic.com/science/2019/10/climate-change-killing-thousands-of-years-indigenous-wisdom/

Kay Vandette, "Climate change could make leafy greens, veggies less available," *Earth*, June 11, 2018. https://www.earth.com/news/climate-change-could-make-leafy-greens-veggies-less-available/

93 Current World Population: https://www.worldometers.info/world-population/

94 "Nature's Dangerous Decline 'Unprecedented'; Species Extinction Rates 'Accelerating'" *Intergovernmental Science-Policy Platform on Biodiversity and Ecosystem Services (IPBES)*, May 22, 2019. https://www.ipbes.net/news/Media-Release-Global-Assessment

95 "Ocean Deoxygenation," *International Union for Conservation of Nature*, December 8, 2019. Marine life, fisheries increasingly threatened as the oceans lose oxygen. Even the smallest fall in oxygen levels, when near already existing thresholds, can create significant issues with far-reaching and complex biological and biogeochemical implications. https://www.iucn.org/theme/marine-and-polar/our-work/climate-change-and-oceans/ocean-deoxygenation

96 Adapted from John Fullerton, "Regenerative Capitalism How Universal Principles And Patterns Will Shape Our New Economy," *Capital Institute*, April 2015. https://capitalinstitute.org/wp-content/uploads/2015/04/2015-Regenerative-Capitalism-4-20-15-final.pdf?mc_cid=236080d2f0&mc_eid=2f41fb9d8d

97 "Richest 1% on target to own two-thirds of all wealth by 2030," *The Guardian*, April 7, 2018. https://www.theguardian.com/business/2018/apr/07/global-inequality-tipping-point-2030

98 Duane Elgin, "Limits to Complexity: Are Bureaucracies Becoming Unmanageable," *The Futurist,* December 1977. https://duaneelgin.com/wp-content/uploads/2014/11/Limits-to-Large-Complex-Systems.pdf

99 "Transitions and Tipping Points in Complex Environmental Systems," A report by the *National Science Foundation Advisory Committee for Environmental Research and Education*, 2009. https://www.nsf.gov/ere/ereweb/ac-ere/nsf6895_ere_report_090809.pdf This is not a specific warning, but rather a general one from 2009: "The world is at a crossroads. The global footprint of humans is such that we are stressing natural and social systems beyond their capacities. We must address these complex environmental challenges and mitigate global scale environmental change or accept likely all-pervasive disruptions. . . . The rate of environmental change is outpacing the ability of institutions and governments to respond effectively."

100 T. Schuur, "Arctic Report Card: Permafrost and the Global Carbon Cycle," *NOAA*, 2019. https://arctic.noaa.gov/Report-Card/Report-Card-2019/ArtMID/7916/ArticleID/844/Permafrost-and-the-Global-Carbon-Cycle

101 "Fighting Wildfires Around the World," *Frontline, Wildfire Defense Systems*, 2019. https://www.frontlinewildfire.com/fighting-wildfires-around-world/

102 Carrying capacity estimates, op. cit.

103 Iliana Paul, "Climate Change and Social Justice," *WEDO*, 2014. https://www.wedo.org/wp-content/uploads/wedo-climate-change-social-justice.pdf?utm_source=newsletter&utm_medium=email&utm_content=http%3A//d31hzlhk6di2h5.cloudfront.net/20161107/ce/11/85/a8/5d76d1fbe015e871ef155f93_386x486.png&utm_campaign=Emma%20Newsletter

104 Dmitry Orlov, *Reinventing Collapse: The Soviet Example and American Prospects*, New Society Publishers, 2008. Also see: Tainter, *The Collapse of Complex Societies*, op. cit.

105 Carrying capacity estimates, op. cit.

106 Op. cit., "Nature's Dangerous Decline 'Unprecedented'; Species Extinction Rates 'Accelerating'" *Intergovernmental Science-Policy Platform on Biodiversity and Ecosystem Services (IPBES),* May 22, 2019. https://www.ipbes.net/news/Media-Release-Global-Assessment

107 Plants are also likely to experience the stress and trauma of the great dying. See Nicoletta Lanese, "Plants 'Scream' in the Face of Stress," *Live Science,* December 6, 2019. https://www.livescience.com/plants-squeal-when-stressed.html

108 My assessment that several billion people may perish in the latter portion of the time frame of this scenario (where the world is not powered by fossil fuels) has been described as hyper-optimistic. Jason Brent (http://www.jgbrent.com/about-the-author.html) considers it likely that many more will die. See his reply to my article, "Existential threats, Earth Voice and the Great Transition," *Millennium Alliance for Humanity and the Biosphere,* MAHB, January 21, 2020. https://mahb.stanford.edu/blog/mahb-dialogue-author-humanist-duane-elgin/ Brent writes: "The collapse of civilization will occur because humanity is in overshoot using the resources of 1.7 Earths and going deeper into overshoot every second due to a growing population (expected to grow by 3.2 billion reaching 10.9 billion by the year 2100 — a 41.5% growth in 80 years) and a growing per capita worldwide usage of resources. Simple math shows that to get out of overshoot the human population would have to be reduced to 4.47 billion. If population were to reach 10.9 billion, that would require a reduction in population of 6.43 billion (10.9-4.47= 6.43) (without considering any reduction due to the per capita increase in the usage of resources) to get out of overshoot. Simple statement — there is zero chance that voluntary population control will achieve that reduction (of 6.3 billion) prior to the collapse of civilization and the deaths of billions."

109 The Great Burning began in 2019. See: Laura Paddison, "2019 Was The Year The World Burned," *HuffPost,* December 27, 2019. https://www.huffpost.com/entry/wildfires-california-amazon-indonesia-climate-change_n_5dcd3f4ee4b0d43931d01baf Also:

At least a billion animals are estimated to have died by 2020 in the bushfires in Australia. Lisa Cox, "A billion animals: some of the species most at risk from Australia's bushfire crisis," *The Guardian,* Jan 13, 2020. Ecologist Chris Dickman has estimated more than a billion animals have died around the country — a figure that excludes fish, frogs, bats and insects. "This is just the tip of the iceberg," James Trezise, a policy analyst at the Australian Conservation Foundation, says. "The number of species and ecosystems that have been severely impacted across their ranges is almost certain to be much higher, especially when factoring in less well-known species of reptiles, amphibians and invertebrates." https://www.theguardian.com/australia-news/2020/jan/14/a-billion-animals-the-australian-species-most-at-risk-from-the-bushfire-crisis

The great burning to come is powerfully summarized in the following video that shows a woman rescuing a badly burnt and wailing koala bear from an Australian bushfire. The marsupial was spotted crossing a road among the flames. A local woman rushed to the koala's aid, wrapping the animal in her shirt and a blanket and pouring water over it. She took the injured animal to a nearby koala hospital. It is truly heartbreaking to watch innocents suffer for reasons not their own — and to realize this is our future unless we respond swiftly. https://www.youtube.com/watch?v=3x8JXQ6RTIU

110 "Wildfires in the Amazon are predicted to worsen, doubling the affected area of an important part of the forest by 2050. The result could be to shift the Amazon from a carbon sink into a net source of carbon dioxide emissions," See story: "Burning of Amazon may get a lot worse," *New Scientist*, January 18, 2020. Also:

Herton Escobar, "Brazil's deforestation is exploding — and 2020 will be worse," *Science Magazine*, November 22, 2019. https://www.sciencemag.org/news/2019/11/brazil-s-deforestation-exploding-and-2020-will-be-worse?utm_campaign=news_daily_2019-11-22&et_rid=510705016&et_cid=3086753

111 Stephen Pune, "California wildfires signal the arrival of a planetary fire age," *The Conversation*, November 1, 2019. https://theconversation.com/california-wildfires-signal-the-arrival-of-a-planetary-fire-age-125972

112 John Pickrell, "Massive Australian blazes will 'reframe our understanding of bushfire,'" *Science Magazine*, November 20, 2019. https://www.sciencemag.org/news/2019/11/massive-australian-blazes-will-reframe-our-understanding-bushfire?utm_campaign=news_daily_2019-11-20&et_rid=510705016&et_cid=3083308 Also: Damien Cave, "Australia Burns Again, and Now Its Biggest City Is Choking," *The New York Times*, December 6, 2019. https://www.nytimes.com/2019/12/06/world/australia/sydney-fires.html

113 Stephen Pyne, "The Planet is Burning," *Aeon*, November 2019. Also:

Stephen Pyne, *Fire: A Brief History* (2019). https://aeon.co/essays/the-planet-is-burning-around-us-is-it-time-to-declare-the-pyrocene

David Wallace-Wells, "In California, Climate Change Has Turned Rainy Season Into Fire Season," *New York Magazine*, November 12, 2018. https://nymag.com/intelligencer/2018/11/the-california-fires-and-the-threat-of-climate-change.html

Edward Helmore, "'Unprecedented': more than 100 Arctic wildfires burn in worst-ever season," *The Guardian*, July 26, 2019. The article describes, "Huge blazes in Greenland, Siberia, and Alaska are producing plumes of smoke that can be seen from space." https://www.theguardian.com/world/2019/jul/26/unprecedented-more-than-100-wildfires-burning-in-the-arctic-in-worst-ever-season

114 Hans Seyle, was a highly regarded endocrinologist known for his studies of the effects of stress on the human body. https://www.azquotes.com/author/13308-Hans_Selye

115 Francis Weller, *The Wild Edge of Sorrow*, North Atlantic Books, 2015. https://www.amazon.com/Wild-Edge-Sorrow-Rituals-Renewal/dp/1583949763

116 Weller, ibid. https://www.amazon.com/Wild-Edge-Sorrow-Rituals-Renewal/dp/1583949763

117 Naomi Shihab Nye, *Words Under the Words: Selected Poems*, 1995. https://poets.org/poem/kindness

118 "Global Cities at Risk from Sea-Level Rise: Google Earth Video," *Climate Central*, 2019. https://sealevel.climatecentral.org/maps/google-earth-video-global-cities-at-risk-from-sea-level-rise Also:

Scott Kulp, et al., "New elevation data triple estimates of global vulnerability to sea-level rise and coastal flooding," *Nature Communications*, October 29, 2019. Some of the earlier projections of population displacement from sea-level rise are probably way too low. Around the world, instead of some 50 million people being forced to move to higher ground over the next 30 years, the oceans will likely rise higher than predicted, with a coastal diaspora at least three times larger; by 2100, the number of climate refugees could surpass 300 million. https://www.nature.com/articles/s41467-019-12808-z Other estimates place the number of climate refugees as high as 2 billion by 2100.

Charles Geisler & Ben Currens, "Impediments to inland resettlement under conditions of accelerated sea-level rise," *Land Use Policy*, March 29, 2017. The authors extrapolate from 2060 to conclude that in the year 2100, 2 billion people — about one-fifth of a world population of 11 billion — could become climate change refugees due to rising ocean levels. https://doi.org/10.1016/j.landusepol.2017.03.029

Blaine Friedlander, "Rising seas could result in 2 billion refugees by 2100," *Cornell Chronicle*, June 19, 2017. http://news.cornell.edu/stories/2017/06/rising-seas-could-result-2-billion-refugees-2100

119 Jennifer Welwood, "The Dakini Speaks," http://jenniferwelwood.com/poetry/the-dakini-speaks/

120 Todd May, "Would Human Extinction Be a Tragedy?" *The New York Times*, December 17, 2018. https://www.nytimes.com/2018/12/17/opinion/human-extinction-climate-change.html

121 Wallace Stevens, *Goodreads*, https://www.goodreads.com/quotes/565035-after-the-final-no-there-comes-a-yes-and

122 Joanna Macy and Chris Johnstone, Active Hope: How to Face the Mess We're in Without Going Crazy, New World Library, 2012.

123 To illustrate the difficulty of meeting the net zero CO_2 emission goals by 2050, see the *World Energy Outlook 2019,* which concludes that the world's CO_2 emissions are set to continue rising for decades unless there is greater ambition on climate change, despite the "profound shifts" already underway in the global energy system. That is one of the key messages from the International Energy Agency's (IEA) https://www.iea.org/reports/world-energy-outlook-2019

124 Of great concern is when cumulative global CO_2 emissions exceed the 1 trillion tons of carbon threshold, which according to the IPCC will raise the Earth's surface temperature to 2°C above the pre-industrial minimum and trigger "dangerous interference" with the Earth's climate system. When will the 1 trillion-ton threshold be exceeded? Estimates are sometime between 2050 and 2055 regardless of which population growth scenario is used. "Global CO_2 emissions forecast to 2100," Roger Andrews, *Euanmearns,* March 7, 2018. http://euanmearns.com/global- CO2-emissions-forecast-to-2100/

125 "Impacts of a 4-degree Celsius Global Warming," *Green Facts,* https://www.greenfacts.org/en/impacts-global-warming/l-2/index.htm Also:

A broad consensus exists that 4°C will happen by the end of the century or before, if no major actions are taken. "Climate change may be escalating so fast it could be 'game over,' scientists warn." A climate range between 4.8C and 7.4C by 2100 emerged from calculations published in the journal, *Science Advances.* https://advances.sciencemag.org/content/2/11/e1501923

Ian Johnston, "Climate change may be escalating so fast it could be 'game over,' scientists warn." *Independent,* November 9, 2016. https://www.independent.co.uk/news/science/climate-change-game-over-global-warming-climate-sensitivity-seven-degrees-a7407881.html

David Wallace-Wells, "U.N. says climate genocide is coming," *New York Magazine,* October 10, 2019. He states that the planet is on a trajectory that "brings us north of four degrees by the end of the century." http://nymag.com/intelligencer/2018/10/un-says-climate-genocide-coming-but-its-worse-than-that.html

Roger Andrews, "Global CO_2 emissions forecast to 2100," Blog by *Euan Mearns,* March 7, 2018. http://euanmearns.com/global-CO2-emissions-forecast-to-2100/

"4 Degrees Hotter, A Climate Action Centre Primer," *Climate Code Red,* February 2011. Melbourne, Australia. https://www.climatecodered.org/2011/02/4-degrees-hotter-adaptation-trap.html The study quotes Professor Kevin Anderson, director of the Tyndall Centre for Climate Change, who "believes only around 10 percent of the planet's population — around half a billion people — will survive if global temperatures rise by 4°C. He said the consequences were "terrifying." "For humanity, it's a matter of life or death," he said. "We will not make all human beings extinct, as a few people with the right sort of resources may put themselves in the right parts of the world and survive. But I think it's extremely unlikely that we wouldn't have mass death at 4°C." In 2009, Professor Hans Joachim Schellbhuber, director of the Potsdam Institute, and one of Europe's most eminent climate scientists, told his audience that at 4°C, population "carrying capacity estimates (are) below 1 billion people." p. 9.

Another estimate of the carrying capacity of the Earth is found in *New Scientist,* November1, 2014, p. 9. Corey Bradshaw and Barry Brook, (op. cit.), suggest that a sustainable human population, given current Western consumption patterns and technologies, would be between 1 and 2 billion people.

126 The researchers used the MIT Integrated Global System Model Water Resource System (IGSM-WRS) to evaluate water resources and needs worldwide. See: "Water Stress to Affect 52% of World's Population by 2050," *Water Footprint Network,* https://waterfootprint.org/en/about-us/news/news/water-stress-affect-52-worlds-population-2050/

127 Op. cit. The United Nations world water development report 2018: nature-based solutions for water. Also:

Claire Bernish, "Water Scarcity Will Make Life Miserable for Nearly 6 Billion People by 2050," *The Mind Unleashed,* March 23, 2018. https://themindunleashed.com/2018/03/water-scarcity-6-billion-2050.html More than 5 billion people could suffer water shortages by 2050 due to climate change, increased demand and polluted supplies, according to a U.N. report on the state of the world's water. Without drastic changes focused on natural solutions, nearly six billion people will be in the grips of a punishing water shortage by 2050.

128 Joseph Hinks, "The World Is Headed for a Food Security Crisis," *TIME* magazine, March 28, 2018. https://time.com/5216532/global-food-security-richard-deverell/

129 Rebecca Chaplin-Kramer, "Global modeling of nature's contributions to people," *Science,* Vol. 366, Issue 6462, October 11, 2019. https://science.sciencemag.org/content/366/6462/255 Also:

Miyo McGinn, "New study pinpoints the places most at risk on a warming planet," *Grist,* October 17, 2019. https://grist.org/article/new-study-pinpoints-the-places-most-at-risk-on-a-warming-planet/

130 Francois Gemenne, "A review of estimates and predictions of people displaced by environmental changes," Global Environmental Change, *in Science Direct,* December 2011. https://www.sciencedirect.com/science/article/abs/pii/S0959378011001403?via%3Dihub

131 Worldometers: https://www.worldometers.info/world-population/

132 See, for example, op. cit., Ishan Daftardar, "Why Bee Extinction Would Mean the End of Humanity," *Science ABC,* July 23, 2015. https://www.scienceabc.com/nature/bee-extinction-means-end-humanity.html

133 "Russia 'meddled in all big social media' around U.S. election," BBC, December 17, 2018. https://www.bbc.com/news/technology-46590890

134 Charles Geisler & Ben Currens, "Impediments to inland resettlement under conditions of accelerated sea-level rise," *Land Use Policy*, March 29, 2017. The authors extrapolate from 2060 to conclude that in the year 2100, 2 billion people—about one-fifth of a world population of 11 billion—could become climate change refugees due to rising ocean levels. https://doi.org/10.1016/j.landusepol.2017.03.029

135 Martin Luther King, Jr. quoted in Stephen B. Oates, *Let the Trumpets Sound: The Life of Martin Luther King, Jr.*, New American Library, 1982.

136 T.S. Eliot, *Four Quartets, Little Gidding*, 1943. https://www.brainyquote.com/quotes/t_s_eliot_109032

137 Drew Dellinger, "Hieroglyphic Stairway," (poem), 2008, https://www.youtube.com/watch?v=XW63UUthwSg

138 Malcolm Margolin, *The Ohlone Way: Indian Life in the San Francisco-Monterey Bay Area*, Berkeley: Heyday Books, 1978.

139 See the wonderful short video by Louie Schwartzberg, *Gratitude*, https://movingart.com/portfolio/gratitude/ Narration written and spoken by Brother David Steindl-Rast. www.MovingArt.com

140 Joseph Campbell, et al., *Changing Images of Man, Center for the Study of Social Policy, Stanford Research Institute*, Menlo Park, California. The study was prepared for the Kettering Foundation, Dayton, Ohio, Contact: URH (489)-2150, May 1974 and subsequently republished with the same title in 1982 by Pergamon Press.

141 Joseph Campbell & Bill Moyers, *The Power of Myth*, Archer, 1988. https://www.goodreads.com/quotes/10442-people-say-that-what-we-re-all-seeking-is-a-meaning

142 Sean D. Kelly, "Waking Up to the Gift of 'Aliveness,'" *The New York Times*, Dec. 25, 2017. https://www.nytimes.com/2017/12/25/opinion/aliveness-waking-up-holidays.html

143 Howard Thurman, https://www.goodreads.com/quotes/6273-don-t-ask-what-the-world-needs-ask-what-makes-you

144 Joanna Macy referenced in Jem Bendell, "Climate despair is inviting people back to life," posted in his blog on deep adaptation, July 12, 2019. https://jembendell.com/

145 Anne Baring, op. cit., p. 83.

146 Ann Baring, op. cit., p. 421.

147 Simone de Beauvoir, See "Brainy Quotes": https://www.brainyquote.com/quotes/simone_de_beauvoir_392724

148 See, for example: https://www.goodreads.com/quotes/tag/mysticism Also: http://www.gardendigest.com/myst1.htm

149 Henry Thoreau, https://www.goodreads.com/quotes/32955-heaven-is-under-our-feet-as-well-as-over-our

150 Predrag Cicovacki, *Albert Schweitzer's Ethical Vision A Sourcebook*, Oxford University Press, February 2, 2009.

151 John Muir, https://www.goodreads.com/quotes/7796963-and-into-the-forest-i-go-to-lose-my-mind

152 Haruki Murakami, https://www.goodreads.com/quotes/448426-not-just-beautiful-though-the-stars-are-like-the

153 Joseph Campbell, https://www.brainyquote.com/quotes/joseph_campbell_387298 There is a subtle and extremely important difference between "consciousness" and "awareness." These two terms are often used interchangeably and yet have very different meanings. Simply stated:

Consciousness reflects — there is always an object of conscious attention.

Awareness is — there is no object of attention, a living presence is aware of itself.

Consciousness refers to the ability to stand back from immersion in thoughts and to witness or observe aspects or elements of life. Consciousness involves two aspects: a knower and that which is known; or an observer and that which is observed; or a watcher and that which is watched. There is a felt distance between consciousness and the object of attention.

Awareness can be described as knowing without an object. *Awareness is aware of itself by its very nature — it just "is."* Awareness is a knowing presence whose nature is awareness. It is simply awareness itself. Awareness is a felt presence, a direct experience of aliveness itself. It is not watching aliveness, it simply the direct experience of aliveness. There is no distance and no separation as this is a singular, felt presence.

How can there be a direct experience of aliveness that extends beyond our physical body? Both physics and wisdom traditions recognize that the entire universe is being uplifted into existence at every moment in an extraordinary process of continuous creation. The regenerative life-force that underlies and uplifts the entire universe at each moment is, by its very nature, aliveness and awareness. *When we become fully one with the direct experience of existence in the moment, we become one with the life force giving rise to the totality of existence.* We recognize ourselves as that life-force as an unbounded, living presence. The life-force of cosmic-scale aliveness is the regenerative force that is upholding the entire universe at every moment — and can be known as a felt experience, as aliveness itself. When our conscious knowing progressively refines to where there is no longer a distance between knowing and what is known, then there is awareness itself.

If we *think* consciousness is essentially a knowing faculty that arises in the brain as a product of intensely complex bio-material interactions, then we create an image of the knowing process that leaves us removed from the direct experience of aliveness and the felt-awareness life-force that sustains the universe moment by moment. Aliveness — as simple, direct awareness — is the home that we seek. *When we are aware of being awareness itself, we are home! At the center of our being is the simplicity of direct experience of being alive, and that experience is awareness itself, and this experience is nothing other than the life-force of cosmic scale creation or "cosmic awareness".*

It is important to allow meditation to rest in the continuity of simple awareness where we release the effort and struggle of returning to an object of attention and simply be with the flow of awareness of what "is." When we ride the direct experience of being awareness itself, we are riding the wave of continuous creation of existence. If we persist in the precise presencing of awareness, it will reveal itself to be the life-force in cosmic scale dance of continuous regeneration. We know, as direct experience, "we are that." We are the undivided life-force of Totality becoming itself and known as the direct experience of being alive.

154 Buddha, https://www.spiritualityandpractice.com/quotes/quotations/view/198/spiritual-quotation

155 Frank Lloyd Wright, https://www.brainyquote.com/quotes/frank_lloyd_wright_107515

156 Florida Scott-Maxwell, *The Measure of My Days*, Penguin Books, 1979. https://www.goodreads.com/author/quotes/550910.Florida_Scott_Maxwell

157 To learn into our time of great transition and beyond, my partner Coleen and I have convened a learning community of roughly several dozen people in the past year. Our collective explorations have been very valuable in grounding the work described in this book.

158 Richard Nelson, *Make Prayers to the Raven*, Chicago: University of Chicago Press, 1983.

159 Luther Standing Bear, quoted in J.E. Brown, "Modes of contemplation through actions: North American Indians." In *Main Currents in Modern Thought*, New York, November-December 1973.

160 Mathew Fox, *Meditations with Meister Eckhart*, Santa Fe, NM: Bear & Co., 1983.

161 See, for example, Coleman Barks, *The Essential Rumi*, San Francisco: Harper San Francisco, 1995.

162 D.T. Suzuki, *Zen and Japanese Culture*, Princeton, NJ: Princeton University Press, 1970.

163 S. N. Maharaj, *I Am That*. Part I (trans., Maurice Frydman), Bombay, India: Chetana, 1973.

164 Lao Tzu, *Tao Te Ching* (trans. Gia-Fu Feng and Jane English), New York: Vintage Books, 1972.

165 E. C. Roehlkepartain, et al., "With their own voices: A global exploration of how today's young people experience and think about spiritual development," *Search Institute*, 2008. www.spiritualdevelopmentcenter.org

166 "Many Americans Mix Multiple Faiths," *Pew Research Center, Religion & Public Life,* December 9, 2009. Mystical experiences shown in third figure, which references 1962 survey reported by Gallup and presented in *Newsweek*, April, 2006 See: https://www.pewforum.org/2009/12/09/many-americans-mix-multiple-faiths/ Also:

Andrew Greely and William McCready, "Are We a Nation of Mystics," in *The New York Times Magazine*, January 26, 1976.

167 "U.S. public becoming less religious," *Pew Research Center*, November 3, 2015. Survey results on regular experiences of "peace and sense of wonder." https://www.pewforum.org/2015/11/03/u-s-public-becoming-less-religious/

168 T. Clarke, et al., "Use of Yoga, Meditation, and Chiropractors Among U.S. Adults Aged 18 and Over," *National Center for Health Statistics*, November 2018. https://www.ncbi.nlm.nih.gov/pubmed/30475686

169 In the spirit of full disclosure, my personal understanding of an ecology of consciousness permeating the universe was developed and documented in a wide-ranging series of scientific experiments over a period nearly three years, from 1972 to 1975, at the Stanford Research Institute (now SRI International), in Menlo Park, California. Although my primary work at the time was as a senior social scientist in the futures group at SRI, for nearly three years, I was a consultant to NASA to explore a wide range of experiments regarding intuitive capacities in the engineering laboratory—often three days a week for two- or three-hour stretches, depending on the experiments at the time, and all with various forms of feedback. Experiments included "remote viewing" of diverse locations and technologies; clairvoyance with a random-number generator; influencing movement of pendulum-clock measured with a laser beam; interacting with a magnetometer whose sensitive probe was immersed in a container filled with liquid helium; standing outside a locked room and pressing on a balance pan scale locked inside; influencing plant growth with comparisons to a controlled group, and more. I dropped out of these fascinating experiments in 1975 when they were taken over by the CIA and declared secret (this research apparently continued for another 20 years according to Freedom of Information Act; see: Hal Puthoff, "CIA-Initiated Remote Viewing Program at Stanford Research Institute," *Journal of Scientific Exploration, Vol. 10*, No. 1, 1996). Based on my experience in these scientific experiments, I learned that:

First, we all have a literal connection with the universe. An empathic connection with the cosmos is not restricted to a gifted few, it is an ordinary part of the functioning of the universe and is accessible to everyone.

Second, our being does not stop at the edge of our skin, but extends into and is inseparable from the universe. We are all connected with the deep ecology of the universe and each of us has the ability to extend our consciousness far beyond the range of our physical senses.

Third, our intuitive connection with the cosmos is easy to overlook. Small twinges of intuitive feelings quickly arise and then pass away. I assumed they were simply part of my bodily experience. Only gradually did I come to appreciate the extent to which I was experiencing my participation in a larger "field" of aliveness.

Fourth, I learned that psi functioning is not about achieving dominance over something (mind over matter), but rather learning to participate with something in a dance of mutual exchange and transformation. This is a two-way process in which both parties are changed by the interaction. In a sentence, domination does not work, but dancing does.

Fifth, at the same time these experiments were showing me how consciousness is an intrinsic property of the universe; they also made me much more skeptical about the need for channeling, crystals, pendulums, pyramids, and other intermediaries to access our intuition. It is important to bring a critical and discerning science to this inquiry.

Sixth, scientific evidence of the existence of psychic functioning has been mounting for decades and is now so overwhelming that the burden of proof has shifted to those who would seek to dismiss its existence. It is time to move beyond the narrow, brain-based view of consciousness because it no longer explains important scientific evidence and it severely limits our thinking about the scope and depth of our connection with the universe.

Seventh, as interesting as psychic or intuitive functioning may be, the much more important question is what it says about the nature of the universe; namely, that it is connected with itself through the tissue of consciousness in non-local ways that transcend relativistic differences.

These experiments made it clear that *we have barely begun to develop a literacy of consciousness using sophisticated technologies to provide feedback* (similar to learning with bio-feedback, but instead with bio-cosmic feedback). These experiments demonstrated that our being does not stop at the edge of our skin, but extends into and is inseparable from the unified universe. A description of selected SRI experiments can be found at:

Russell Targ, Phyllis Cole, and Harold Puthoff, "Development of Techniques to Enhance Man/Machine Communication," *Stanford Research Institute*, Menlo Park, California, prepared for NASA, contract 953653 Under NAS7-100, June 1974. Also:

Harold Puthoff and Russell Targ, "A Perceptual Channel for Information Transfer Over Kilometer Distances," published in the *Proceedings of the I.E.E.E. (Institute of Electrical and Electronics Engineers)*, vol. 64, no. 3, March 1976.

R. Targ and H. Puthoff, *Mind-Reach: Scientists Look at Psychic Ability*, Delacorte Press/Eleaonor Friede, 1977.

170 Duane Elgin, *The Living Universe*, op., cit. Another way to consider the issue of aliveness is to explore the operating characteristics of biological systems and see whether the universe exhibits similar capacities. Generally, a system must include at least four key capacities to be considered living: 1) *Metabolism* — the ability to break matter down as well as to synthesize

it. From its formation, the universe has been synthesizing simple matter (helium and hydrogen) and converting it through supernova into carbon, nitrogen, oxygen and sulfur—essential constituents from which we are made. 2) *Self-regulation*—the ability to maintain stability in its operation. The universe has endured and evolved over billions of years as a unified system that produces self-organizing systems at every scale, from atomic to galactic, that can persist for billions of years. 3) *Reproduction*—the ability to create copies. A number of cosmologists theorize that on the other side of black holes are white holes giving birth to new cosmic systems. 4) *Adaptation*—the ability to evolve and fit into changing environments. The universe has evolved over billions of years to produce systems of increasing complexity and coherence woven together into a self-consistent whole. Because these four criteria are found, not only in plants and animals, but also in the functioning of the universe, it seems valid to describe the universe as a unique kind of living system.

171 Albert Einstein's famous quote was written in 1950 in a letter to Robert S. Marcus, who was distraught over the death of his young son from polio. Originally written in German, it was then translated into English and it is the English version that has been widely distributed. However, the original version in German reveals more accurately Einstein's intended meaning. See: https://www.thymindoman.com/einsteins-misquote-on-the-illusion-of-feeling-separate-from-the-whole/

172 Clara Moskowitz, "What's 96 Percent of the Universe Made Of? Astronomers Don't Know," *Space.com*, May 12, 2011. https://www.space.com/11642-dark-matter-dark-energy-4-percent-universe-panek.html

173 Brian Swimme, *The Hidden Heart of the Cosmos*, Orbis Books, May 1996. https://www.amazon.com/Hidden-Heart-Cosmos-Humanity-Ecology/dp/1626983437

174 Phillip Goff, "Is the Universe a Conscious Mind?" in *Aeon*, 2019. https://aeon.co/essays/cosmopsychism-explains-why-the-universe-is-fine-tuned-for-life. Physicist and cosmologist Freeman Dyson has written, "It appears that mind, as manifested by the capacity to make choices, is to some extent inherent in every electron."

175 See, for example, the classic book by Richard Bucke, *Cosmic Consciousness*, 1901. ISBN 978-0-486-47190-7. https://www.penguinrandomhouse.ca/books/321631/cosmic-consciousness-by-richard-maurice-bucke/9780140193374

176 Max Planck, Interview in *The Observer*, January 25, 1931. https://en.wikiquote.org/wiki/Max_Planck

177 John Gribbin, *In the Beginning: The Birth of the Living Universe*, New York: Little Brown, 1993.

Also see: David Shiga, "Could black holes be portals to other universes?" *New Scientist*, April 27, 2007.

178 Thomas Berry, *The Dream of the Earth*, Sierra Club Books, 1988.

179 Robert Bly (trans.), *The Kabir Book,* Boston: Beacon Press, 1977, p. 11.

180 Cynthia Bourgeault, *The Wisdom Way of Knowing,* Jossey-Bass, 2003, p. 49. https://inwardoutward.org/aliveness-sep-22-2021/

181 Saint Teresa of Avila, *Brainy Quote.* https://www.brainyquote.com/quotes/saint_teresa_of_avila_105360

182 See Dziuban's website: www.PeterDziuban.com

183 Peter Dziuban, "The Meaning of Life Is Alive," *Excellence Reporter,* November 26, 2017. https://excellencereporter.com/2017/11/26/peter-dziuban-the-meaning-of-life-is-alive/

184 See Carl Sagan's testimony: https://www.youtube.com/watch?v=Wp-WiNXH6hI

185 Henri Nouwen, *The Way of the Heart: Connecting with God through Prayer, Wisdom, and Silence,* Harper Collins, 1981.

186 Ted MacDonald & Lisa Hymas, "How broadcast TV networks covered climate change in 2018," *Media Matters,* March 11, 2019. https://www.mediamatters.org/donald-trump/how-broadcast-tv-networks-covered-climate-change-2018

187 Ted MacDonald, "How broadcast TV networks covered climate change in 2020," *Media Matters,* March 10, 2021. https://www.mediamatters.org/broadcast-networks/how-broadcast-tv-networks-covered-climate-change-2020

188 Gene Youngblood, "The Mass Media and the Future of Desire," *The CoEvolution Quarterly* Sausalito, CA:Winter 1977/78.

189 Martin Luther King, Jr., quoted in Stephen B. Oates, *Let the Trumpets Sound: The Life of Martin Luther King, Jr.,* New American Library, 1982.

190 In the U.S., the rights of the public are profound when it comes to use of the airwaves for both radio and television. These rights are established in the Bill of Rights and Constitutional law. The First Amendment in the Bill of Rights states that: *"Congress shall make no law . . . abridging the freedom of speech . . . or the right of people to peaceably assemble, and to petition the Government for a redress of grievances."* In other words, no law will be passed that limits the right of citizens to assemble peacefully, speak freely, and petition the government for redress of grievances. This is exactly what is involved in an electronic town meeting in the modern era: Citizens assemble peacefully. They speak freely. And, if there is a working consensus, they can directly petition the government asking for redress — or for setting matters right or establishing appropriate remedies.

Turning from Constitutional law to media law in the U.S., we find that **the public at the "local level" is the owner of the airwaves used by television broadcasters. The local level is the scope of the media footprint of broadcasters, which is generally the metropolitan scale.** Even if broadcasters use the internet to deliver much of their own programming, if they also use the airwaves, they still have a strict legal obligation "to serve the public interest, convenience, and necessity."

Nearly a century ago, the Radio Act of 1927 established the basic rules for operation using the public's airwaves, stating that: *"broadcast stations are not given these great privileges by the United States Government for the primary benefit of advertisers. Such benefit as is derived by advertisers must be incidental and entirely secondary to the interest of the public."* The Commission further stated that: *"The emphasis must be first and foremost on the interest, convenience and necessity of the listening public, and not on the interest, convenience, or necessity of the individual broadcaster or advertiser."*

A Federal Appeals Court clarified the role of citizens in 1966, saying: *"Under our system, the interests of the public are dominant. . . . Hence, individual citizens and the communities they compose owe a duty to themselves and their peers to take an active interest in the scope and quality of television service the stations and networks provide. . . . Nor need the public feel that in taking a hand in broadcasting they are unduly interfering in the private business affairs of others. On the contrary, their interest in television programming is direct and their responsibilities important. They are the owners of the channels of television—indeed, of all broadcasting."* [emphasis added]

A 1969 Supreme Court decision further clarified the responsibilities of broadcasters. The court ruled that: *"It is the right of the viewers and listeners, not the right of the broadcasters, which is paramount."* [emphasis added] The Communications Act of 1934 was updated by the U.S. Congress in 1996. The resulting *Telecommunications Act* is more than 300 pages long and, throughout, it affirms the principle that the airwaves should be used *"to serve the public interest, convenience, and necessity."* Television broadcasters have no property rights with regard to using the airwaves; they have the privilege of using the airwaves only as long as they serve the public interest, convenience, and necessity. [emphasis added]

Importantly, we have moved beyond a time of serving the "public interest." Given that local communities are threatened by climate change and the viability of the entire planet, *we have moved to a much higher standard for broadcasters; namely that they serve the "public interest" and the "public necessity."* [emphasis added]

In practical terms, this means that if the local public (the metropolitan scale of the broadcaster's media footprint) asks for a reasonable amount of airtime to be devoted to the climate challenge (which threatens a local community, as well as the entire Earth), then the public should expect the support of the government (the Federal Communications Commission) to uphold such requests that clearly serve the public interest and necessity.

Similarly, *if the public requests airtime for electronic town meetings to consider threats such as climate change, these requests for using the airwaves (which we citizens own) are entirely legitimate and are grounded in both Constitutional law and in nearly a century of Federal law.*

191 Duane Elgin and Peter Russell on "Pete and Duane's Window," *Take Back the Airwaves part 2*, January 19, 2011. https://www.youtube.com/watch?v=a53hL5Z1WHE&feature=youtu.be

192 "Number of Olympic Games TV viewers worldwide from 2002 to 2016," *Statista*, 2020. https://www.statista.com/statistics/287966/olympic-games-tv-viewership-worldwide/

193 Regarding access to television: "For the first time, more than half of the world's population with TV sets are now within reach of a digital TV signal. The figure stands at approximately 55 percent as of 2012, compared to just 30 percent in 2008, according to *ITU's* annual "Measuring the Information Society, 2013." Also:

Tom Butts, "The State of Television, Worldwide," *TV Technology*, December 6, 2013. https://www.tvtechnology.com/miscellaneous/the-state-of-television-worldwide With regard to TV households: Global digital penetration climbed from 40.4 percent of TV households at end of 2010 to 74.6 percent by end of 2015, according to the latest edition of the *Digital TV World Databook*. About 584 million digital TV homes were added in 138 countries between 2010 and 2015. This doubled the digital TV household total to 1,170 million.

According to *Digital TV Research*, "Three Quarters of global TV households are now digital," May 12, 2016 https://www.digitaltvnews.net/?p=27448

In 2002, 1.12 billion households—about three quarters of humanity—owned at least one television set. See: http://www.worldwatch.org/node/810

The number of TV households worldwide is projected to increase from 1.74 billion in 2023, up from 1.63 billion in 2017.

"Number of TV households worldwide from 2010 to 2018," *Statista*, December 4, 2019. https://www.statista.com/statistics/268695/number-of-tv-households-worldwide/

As further context: In July 2012: The world had 7 billion people, and they lived in 1.9 billion households, which on average have 3.68 people in each. Of those 1.9bn households, only 1.4bn households have a TV, let alone the internet. https://www.theguardian.com/media/blog/2012/jul/27/4-billion-olympic-opening-ceremony

194 "World Internet Users and 2019 Population Stats," Miniwatts Marketing Group, October 4, 2019. https://www.internetworldstats.com/stats.htm Some of the key takeaways from their Global Digital Report 2019 include: The number of internet users worldwide in 2019 is 4.388 billion, up 9.1 percent year-on-year. The number of social media users worldwide in 2019 is 3.484 billion, up 9 percent year-on-year. The number of mobile phone users in 2019 is 5.112 billion, up 2 percent year-on-year. See: https://hootsuite.com/pages/digital-in-2019

195 A. W. Geiger, "Key Findings about the online news landscape in America," *Pew Research Center*, September 11, 2019. https://www.pewresearch.org/fact-tank/2019/09/11/key-findings-about-the-online-news-landscape-in-america/ Perspective on the U.S. experience: a Pew Research study found that in 2019, 49 percent of Americans get their news often from television, 33 percent from online websites, 26 percent from radio, 20 percent from social media, and 16 percent from print newspapers.

196 Maya Angelou, *Letter to My Daughter*, Random House, 2008.

197 Toni Morrison, "2004 Wellesley College commencement address," published in *Take This Advice: The Best Graduation Speeches Ever Given*, Simon & Schuster, 2005.

198 Christopher Bache, *Dark Night, Early Dawn: Steps to a Deep Ecology of Mind*, New York: SUNY Press, 2000.

199 See, for example: Joseph V. Montville, "Psychoanalytic Enlightenment and the Greening of Diplomacy," *Journal of the American Psychoanalytic Association*, Vol. 37, No. 2, 1989. Also:

Roger Walsh, *Staying Alive: The Psychology of Human Survival*, Boulder Colorado: New Science Library, 1984.

200 Martin Luther King, Jr., https://www.brainyquote.com/quotes/martin_luther_king_jr_101309

201 Alan Paton, https://www.azquotes.com/author/11383-Alan_Paton

202 See, for example: Dana Meadows, et. al., *Beyond the Limits,* Chelsea Green Publishing Co., 1992.

203 Tatiana Schlossberg [An interview with Narasimha Rao, professor at Yale], "Taking a Different Approach to Fighting Climate Change," *The New York Times*, November 7, 2019. https://www.nytimes.com/2019/11/07/climate/narasimha-rao-climate-change.html Also:

Environmental and Climate Justice Program, *NAACP*, https://www.naacp.org/environmental-climate-justice-about/

"Climate justice," *Wikipedia*, "A fundamental proposition of climate justice is that those who are least responsible for climate change suffer its gravest consequences." https://en.wikipedia.org/wiki/Climate_justice

204 Pedro Conceição, et al, "Human Development Report: Beyond income, beyond averages, beyond today: Inequalities in human development in the 21st century," *UNDP*, 2019 http://hdr.undp.org/sites/default/files/hdr2019.pdf

205 "Forced from Home: Climate-fueled displacement," *Oxfam Media Briefing*, December 2, 2019. https://oxfamilibrary.openrepository.com/bitstream/handle/10546/620914/mb-climate-displacement-cop25-021219-en.pdf "Countries that contribute the least to greenhouse gas emissions will likely continue to experience the greatest consequences due to climate change. The greatest impact of climate change will occur in poor countries." Also:

Barry Levy, et. al., "Climate Change and Collective Violence," *Annual Review of Public Health*, January 11, 2017. doi: 10.1146/annurev-publhealt h-031816-044232

"Environmental & Climate Justice," *NAACP*, 2019. https://www.naacp.org/issues/environ mental-justice/

206 The soul of the universe from the perspective of a feminine archetype has been beautifully developed by the scholar Anne Baring. See her magnificent book, *The Dream of the Cosmos*, Archive Publishing, 2013. Free download at: https://all-med.net/pdf/the-dream-of-the-cosmos/

207 The evolution from an "Earth Goddess" perspective to a "Sky God" perspective to the rise of the "Cosmic Goddess" is explored in my book, *Awakening Earth*, op. cit, 1993. https://duaneelgin.com/wp-content/uploads/2016/03/AWAKENING-EARTH-e-book-2.0.pdf

208 Desmond Tutu quoted in Terry Tempest Williams, *Two Words*, Orion, Great Barrington, MA, Winter 1999.

209 These examples were drawn, in part, from: Emily Mitchell, "The Decade of Atonement," *Index on Censorship*, May/June 1998, London (and reprinted in the *Utne Reader*, March-April 1999).

210 John Bond, "Aussie Apology," *Yes! A Journal of Positive Futures*, Bainbridge Island: WA, Fall 1998.

211 Ibid.

212 Eric Yamamoto, *Interracial Justice: Conflict and Reconciliation in Post-Civil Rights America*, New York University Press, 1999.

213 Alexander, Christopher (1979). The Timeless Way of Building. Oxford University Press. ISBN 978-0-19-502402-9.

214 Ecovillage; see: https://en.wikipedia.org/wiki/Ecovillage Also:

"Global Ecovillage Network": https://ecovillage.org/
https://www.ic.org/directory/ecovillages/
In the United States: https://www.transitionus.org/transition-towns

EcoDistricts. https://ecodistricts.org/ "Within every neighborhood (or district) lies the opportunity to design truly innovative, scalable solutions to some of the biggest challenges facing city makers today: income, education, and health disparities; blight and ecological degradation; the growing threat of climate change; and rapid urban growth. EcoDistricts is advancing a new model of urban development to empower just, sustainable, and resilient neighborhoods. [EcoDistricts are a] . . . collaborative, holistic, neighborhood-scale approach to community design to achieve rigorous, meaningful performance outcomes that matter to people and planet."

215 Transition Towns refer to grassroot community projects that aim to increase self-sufficiency to reduce the potential effects of peak oil, climate destruction, and economic instability. See: https://en.wikipedia.org/wiki/Transition_town Also:

https://transitionnetwork.org/ Here is a list of transition "hubs" around the world: https://transitionnetwork.org/transition-near-me/hubs/

216 See: https://en.wikipedia.org/wiki/Sustainable_city See how sustainable cities fit within the United Nations "sustainable development goals." https://www.un.org/sustainabledevelopment/cities/

Also, for European sustainable cities, see: http://www.sustainablecities.eu/

217 Eco-civilizations: See: https://en.wikipedia.org/wiki/Ecological_civilization Pressure is building to take radical action to de-carbonize the economy as the window for mitigation is closing. A substantial emission reduction is needed before 2030 if global warming is to be kept below 2°C. Several countries have started to change policy and are about to transition into eco-civilizations, changes supported by benefits beyond mitigating climate change (e.g. health benefits). China is a world leader. Also:

"Eco-civilization: China's blueprint for a new era." https://thediplomat.com/2015/09/chinas-new-blueprint-for-an-ecological-civilization/

https://www.creavis.com/sites/creavis/en/creavis/portfolio-development/corporate-foresight/pages/deep-de-carbonization.aspx

218 Alan AtKisson, *Life Beyond Growth,* AtKisson Group, Stockholm, Sweden, 2012. https://wachstumimwandel.at/wp-content/uploads/presentations/AtKisson_GrowthinTransition_Vienna_Oct2012_v1.pdf Even these estimates may underestimate the cost of climate change. Also:

Naomi Oreskes and Nicholas Stern, "Climate Change Will Cost Us Even More Than We Think," *The New York Times,* Oct. 23, 2019. https://www.nytimes.com/2019/10/23/opinion/climate-change-costs.html

219 See, for example, the Swedish word "*lagom,*" which means "just the right amount," "in balance," "perfect-simple." https://en.wikipedia.org/wiki/Lagom

220 Arnold Toynbee, *A Study of History,* (Abridgement of Vol's I-VI, by D.C. Somervell), New York: Oxford University Press, 1947, p. 198.

221 Robert McNamara, former president of the World Bank, defined "absolute poverty" as: "a condition of life so characterized by malnutrition, illiteracy, disease, high infant mortality, and low life expectancy as to be beneath any reasonable definition of human decency."

222 For various definitions, see, Elgin, *Voluntary Simplicity*, op. cit., (first edition, 1981), p. 29. https://www.amazon.com/Voluntary-Simplicity-Toward-Outwardly-Inwardly/dp/0061779261

223 Buckminster Fuller describes this process as "ephemeralization." However, unlike Toynbee, Fuller emphasized designing material systems to do more with less rather than the co-evolution of matter and consciousness. See, for example, his book, *Critical Path*, New York: St. Martin's Press, 1981.

224 Matthew Fox, *Creation Spirituality,* San Francisco: HarperSan Francisco, 1991.

225 Francis J. Flynn, "Where Americans Find Meaning in Life," *Pew Research Center*, November 20, 2018, https://www.pewforum.org/2018/11/20/where-americans-find-meaning-in-life/ Also:

"Research: Can Money Buy Happiness?," *Stanford Business*, September 25, 2013. https://www.gsb.stanford.edu/insights/research-can-money-buy-happiness

"Can Money Buy You Happiness?" Andrew Blackman, *Wall Street Journal*, November 10, 2014. Research shows how life experiences give us more lasting pleasure than material things is found here: https://www.wsj.com/articles/can-money-buy-happiness-heres-what-science-has-to-say-1415569538

Sean D. Kelly, "Waking Up to the Gift of 'Aliveness,'" *New York Times*, December 25, 2017. https://www.nytimes.com/2017/12/25/opinion/aliveness-waking-up-holidays.html

226 Op. cit., "Can Money Buy You Happiness?" Andrew Blackman.

227 Ronald Inglehart, Roberto Foa, et. al. "Development, Freedom, and Rising Happiness: A Global Perspective (1981–2007), July, 2008. Association for Psychological Science, Vol. 3, No., 4, 2008. Find in PubMed: https://doi.org/10.1111/j.1745-6924.2008.00078.x Also:

Ronald Inglehart, "Changing Values among Western Publics from 1970 to 2006," *West European Politics*, January-March 2008. https://www.tandfonline.com/doi/abs/10.1080/01402380701834747

228 Ralph Waldo Emerson. See: https://philosiblog.com/2013/06/10/the-only-true-gift-is-a-portion-of-yourself/

229 Roger Walsh, "Contributing Effectively In Times of Crisis," November 16, 2020. https://www.whatisemerging.com/opinions/contributing-effectively-in-times-of-crisis

Made in the USA
Monee, IL
06 June 2023